ArcGIS® 9

Geocoding in ArcGIS®

DATA CREDITS

Burglary Map: GDT and Vexcel Corp.

Yellowstone Fire Data: National Park Service

Quick-Start Tutorial Data: Atlanta, Georgia

WRITER
Scott Crosier

Contents

Introduction

1

IN THIS CHAPTER

- **Applications of geocoding**
- **Geocoding and ArcGIS**
- **Tips on learning the geocoding process**

Geocoding is the process of assigning a location, usually in the form of coordinate values, to an address by comparing the descriptive location elements in the address to those present in the reference material. Addresses come in many forms, ranging from the common address format of house number followed by the street name and succeeding information to other location descriptions such as postal zone or census tract. In essence, an address includes any type of information that distinguishes a place.

This chapter begins by illustrating many of the applications of geocoding. Next, a brief overview of the tools and software available with ESRI® ArcGIS® is given. Finally, tips on applying these tools to the geocoding process will be presented.

Applications of geocoding

Converting an address to a specific point location on a map—commonly referred to as geocoding—can serve as a valuable tool to suit your specific needs. From simple data analysis to business and customer management to distribution techniques, there is a wide range of applications where geocoding can be used.

Address data analysis

With geocoded addresses, you can spatially display the address locations and begin to recognize patterns within the information. This can be done by simply looking at the information or by using some of the analysis tools available with ArcGIS. You can also display your address information based on certain parameters, allowing you to further analyze the information.

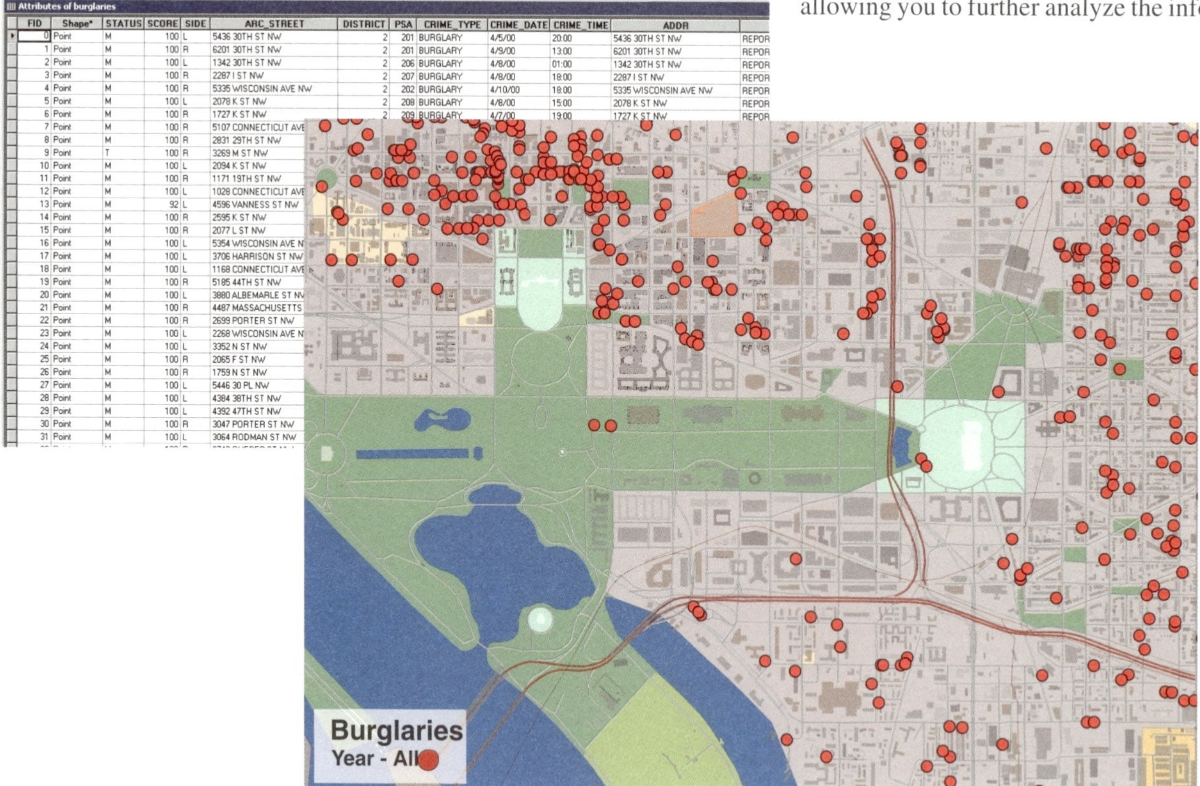
Attributes of burglaries

FID	Shape*	STATUS	SCORE	SIDE	ARC_STREET	DISTRICT	PSA	CRIME_TYPE	CRIME_DATE	CRIME_TIME	ADDR	
0	Point	M	100	L	5436 30TH ST NW	2	201	BURGLARY	4/5/00	20:00	5436 30TH ST NW	REPOR
1	Point	M	100	R	6201 30TH ST NW	2	201	BURGLARY	4/9/00	13:00	6201 30TH ST NW	REPOR
2	Point	M	100	L	1342 30TH ST NW	2	206	BURGLARY	4/8/00	01:00	1342 30TH ST NW	REPOR
3	Point	M	100	R	2287 I ST NW	2	207	BURGLARY	4/8/00	18:00	2287 I ST NW	REPOR
4	Point	M	100	R	5335 WISCONSIN AVE NW	2	202	BURGLARY	4/10/00	18:00	5335 WISCONSIN AVE NW	REPOR
5	Point	M	100	L	2078 K ST NW	2	208	BURGLARY	4/8/00	15:00	2078 K ST NW	REPOR
6	Point	M	100	R	1727 K ST NW	2	209	BURGLARY	4/7/00	19:00	1727 K ST NW	REPOR
7	Point	M	100	R	5107 CONNECTICUT AVE							
8	Point	M	100	R	2831 29TH ST NW							
9	Point	T	100	R	3269 M ST NW							
10	Point	M	100	L	2094 K ST NW							
11	Point	M	100	R	1171 19TH ST NW							
12	Point	M	100	L	1028 CONNECTICUT AVE							
13	Point	M	92	L	4596 VANNESS ST NW							
14	Point	M	100	R	2595 K ST NW							
15	Point	M	100	R	2077 L ST NW							
16	Point	M	100	L	5354 WISCONSIN AVE N							
17	Point	M	100	L	3706 HARRISON ST NW							
18	Point	M	100	L	1168 CONNECTICUT AVE							
19	Point	M	100	R	5185 44TH ST NW							
20	Point	M	100	L	3880 ALBEMARLE ST NV							
21	Point	M	100	R	4487 MASSACHUSETTS							
22	Point	M	100	R	2699 PORTER ST NW							
23	Point	M	100	L	2268 WISCONSIN AVE N							
24	Point	M	100	L	3352 N ST NW							
25	Point	M	100	R	2065 F ST NW							
26	Point	M	100	R	1759 N ST NW							
27	Point	M	100	L	5446 30 PL NW							
28	Point	M	100	L	4384 38TH ST NW							
29	Point	M	100	L	4392 47TH ST NW							
30	Point	M	100	R	3047 PORTER ST NW							
31	Point	M	100	L	3064 RODMAN ST NW							

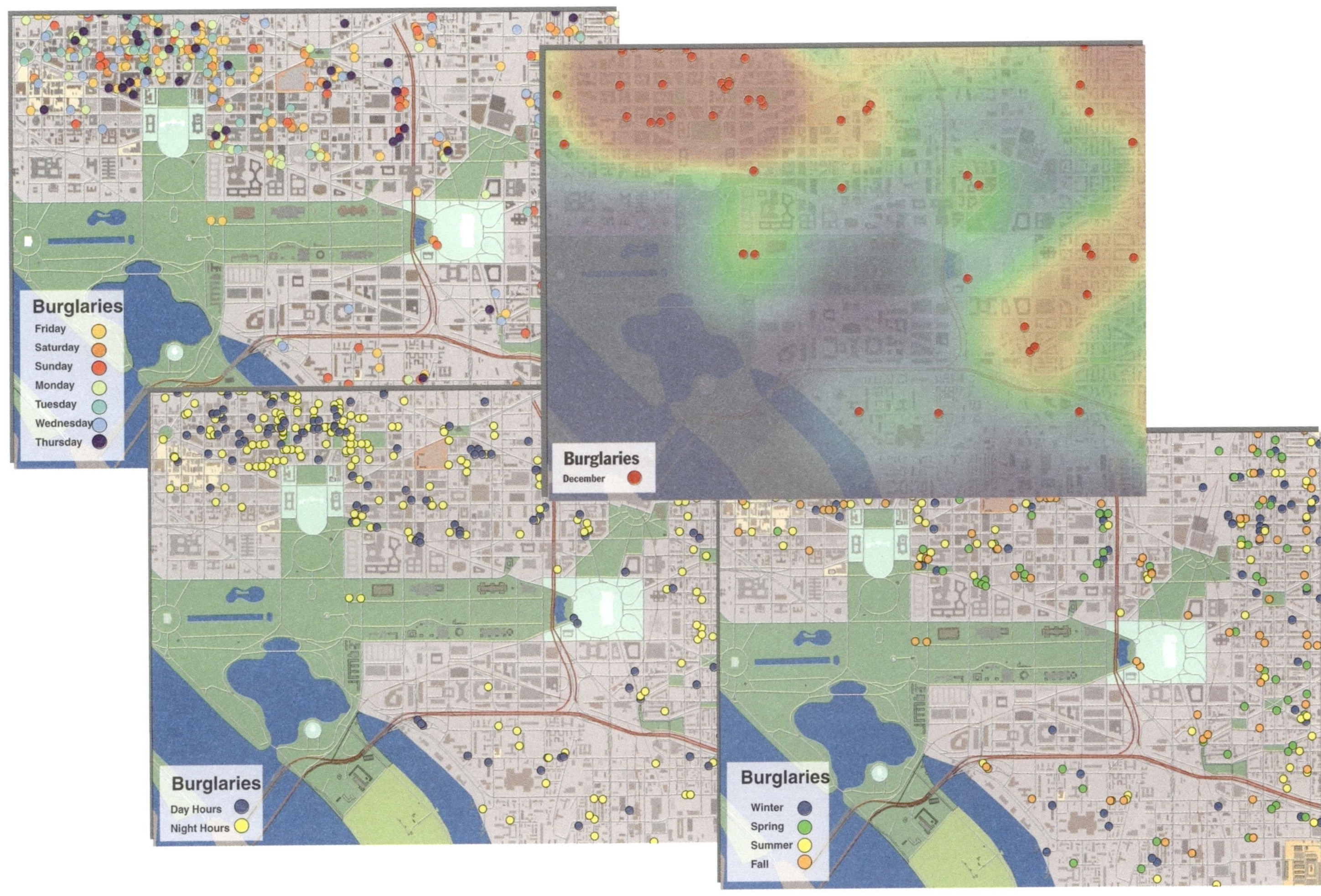

As illustrated on the facing page, the annual record of burglaries was initially created by geocoding a database table of burglaries that consisted of an address for each. The illustrations above show how the geocoded addresses were presented according to time, season, and day of the week to assist in crime prevention planning. Additional analysis tools available in ArcGIS could be used to further analyze the information to identify patterns.

Base data provided by GDT and Vexcel Corp.

Customer data management

Geocoding acts as a crucial part of customer data management. Nearly every organization maintains address information for each customer or client. This is usually in a tabular format, containing the customer name, address, buying habits, and any other information you have collected. Geocoding allows you to take your customer information and create a map of their locations. Using a variety of related applications, you can use this information in many ways, from establishing marketing strategies to targeting specific clusters of customers to producing route maps and directions. The geocoded locations of your customers can be invaluable data.

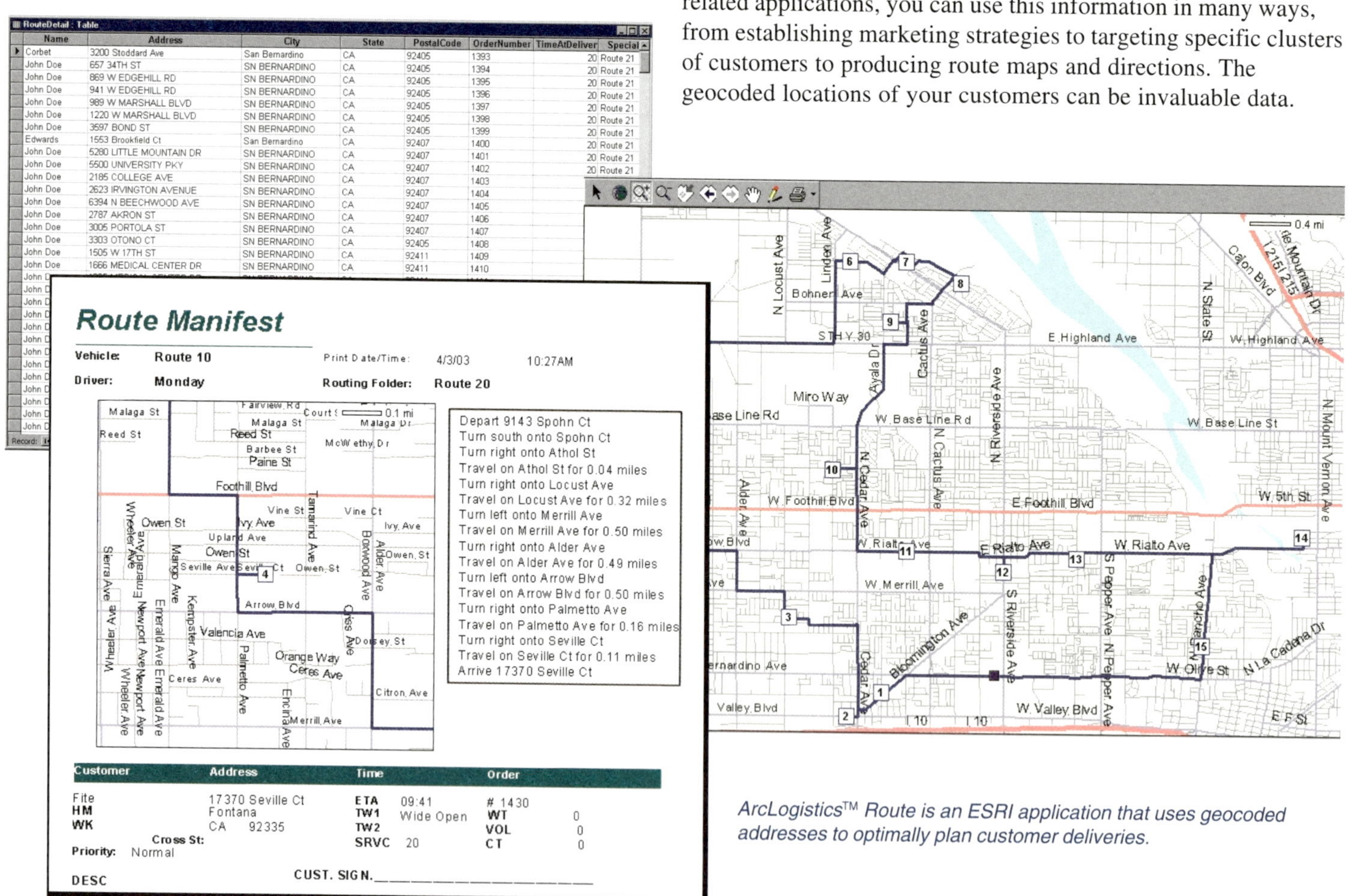

RouteDetail : Table

Name	Address	City	State	PostalCode	OrderNumber	TimeAtDeliver	Special
Corbet	3200 Stoddard Ave	San Bernardino	CA	92405	1393	20	Route 21
John Doe	657 34TH ST	SN BERNARDINO	CA	92405	1394	20	Route 21
John Doe	869 W EDGEHILL RD	SN BERNARDINO	CA	92405	1395	20	Route 21
John Doe	941 W EDGEHILL RD	SN BERNARDINO	CA	92405	1396	20	Route 21
John Doe	989 W MARSHALL BLVD	SN BERNARDINO	CA	92405	1397	20	Route 21
John Doe	1220 W MARSHALL BLVD	SN BERNARDINO	CA	92405	1398	20	Route 21
John Doe	3597 BOND ST	SN BERNARDINO	CA	92405	1399	20	Route 21
Edwards	1553 Brookfield Ct	San Bernardino	CA	92407	1400	20	Route 21
John Doe	5280 LITTLE MOUNTAIN DR	SN BERNARDINO	CA	92407	1401	20	Route 21
John Doe	5500 UNIVERSITY PKY	SN BERNARDINO	CA	92407	1402	20	Route 21
John Doe	2185 COLLEGE AVE	SN BERNARDINO	CA	92407	1403		
John Doe	2623 IRVINGTON AVENUE	SN BERNARDINO	CA	92407	1404		
John Doe	6394 N BEECHWOOD AVE	SN BERNARDINO	CA	92407	1405		
John Doe	2787 AKRON ST	SN BERNARDINO	CA	92407	1406		
John Doe	3005 PORTOLA ST	SN BERNARDINO	CA	92407	1407		
John Doe	3303 OTONO CT	SN BERNARDINO	CA	92405	1408		
John Doe	1505 W 17TH ST	SN BERNARDINO	CA	92411	1409		
John Doe	1666 MEDICAL CENTER DR	SN BERNARDINO	CA	92411	1410		

Route Manifest

Vehicle: Route 10 — Print Date/Time: 4/3/03 10:27AM

Driver: Monday — Routing Folder: Route 20

Depart 9143 Spohn Ct
Turn south onto Spohn Ct
Turn right onto Athol St
Travel on Athol St for 0.04 miles
Turn right onto Locust Ave
Travel on Locust Ave for 0.32 miles
Turn left onto Merrill Ave
Travel on Merrill Ave for 0.50 miles
Turn right onto Alder Ave
Travel on Alder Ave for 0.49 miles
Turn left onto Arrow Blvd
Travel on Arrow Blvd for 0.50 miles
Turn right onto Palmetto Ave
Travel on Palmetto Ave for 0.16 miles
Turn right onto Seville Ct
Travel on Seville Ct for 0.11 miles
Arrive 17370 Seville Ct

Customer	Address	Time		Order	
Fite	17370 Seville Ct	ETA	09:41	# 1430	
HM	Fontana	TW1	Wide Open	WT	0
WK	CA 92335	TW2		VOL	0
	Cross St:	SRVC	20	CT	0

Priority: Normal

DESC — CUST. SIGN.

ArcLogistics™ Route is an ESRI application that uses geocoded addresses to optimally plan customer deliveries.

Distributed geocoding applications

There is a wide range of methods that you can use to share your geocoding functionality. These include everything from collecting the needed material and sharing via a compressed file or compact disc to developing an online application, allowing users to do geocoding over the Internet.

Many real estate firms have found advantages in distributing information about available real estate via the Internet. By combining the database of available homes and ArcGIS Web services, the spatial and nonspatial information about a home can be distributed to a wide audience.

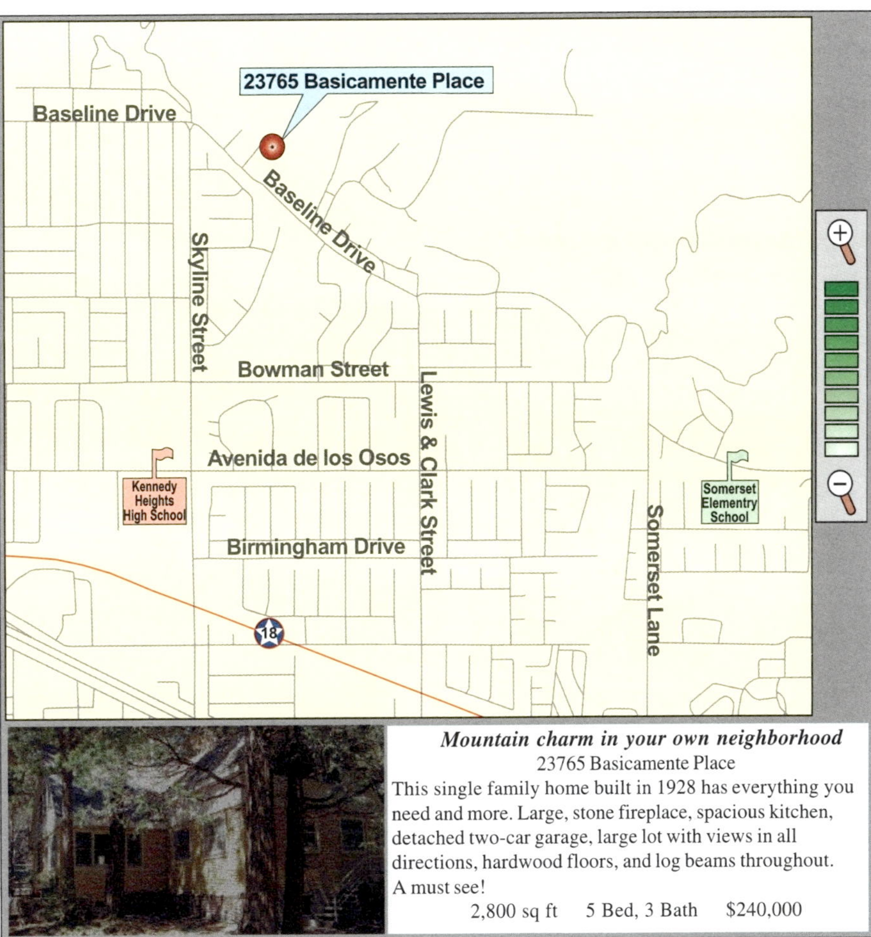

Geocoding and ArcGIS

The ArcGIS system consists of several integrated tools and applications. These are ArcCatalog™, ArcMap™, toolboxes, ArcSDE®, and ArcGIS Server. As you progress through the geocoding process, you may take advantage of each one of these applications or tools.

ArcCatalog

The first of these applications that you will take advantage of is ArcCatalog. ArcCatalog helps you organize all of your geographic information system (GIS) data, including the geocoding applications that you will create in the geocoding process. There are many advantages to using ArcCatalog for GIS data storage. Following is a brief description of some of the geocoding-specific features of ArcCatalog. For a more complete description of ArcCatalog, see *Using ArcCatalog*.

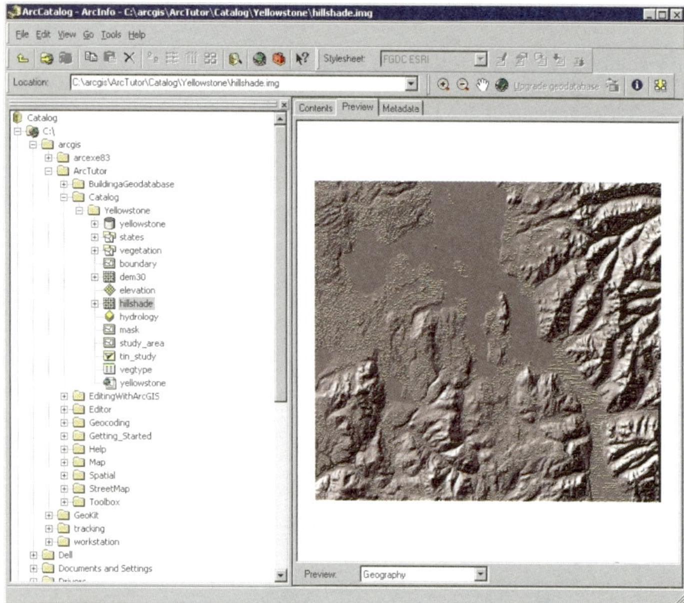

While ArcCatalog appears similar to many other file exploration applications, it allows you to perform many other data maintenance and manipulation processes.

Organizing reference and address data

As feature classes and tables, your reference and address material is easily organized and managed through ArcCatalog. Specific icons related to the GIS-related data types, a catalog tree, and a variety of methods of visualizing your data provide an integrated method of simplifying the organization process.

Creating address locators

Creating data-specific tools for geocoding—referred to as address locators—is done within ArcCatalog. Through a series of easy-to-follow dialog boxes, you can create the address locator to best suit your specific geocoding tasks.

Geocoding tables of addresses

You may wish to geocode tables of addresses directly in ArcCatalog. This creates a feature class that can then be added directly into ArcMap.

Publishing address locators to share with other users

ArcCatalog also provides you with the tools to collect and distribute information and applications related to geocoding. Distribution of files through ArcSDE and ArcGIS Server is done through ArcCatalog.

ArcMap

The second application that will be used heavily in the geocoding process is ArcMap. ArcMap is the central application in ArcGIS Desktop. It is used to perform most of the map-based tasks including cartography, map analysis, and editing. ArcMap provides a wide range of components that facilitate GIS processes. For a more detailed description of ArcMap software's tools and functionality, see *Using ArcMap*.

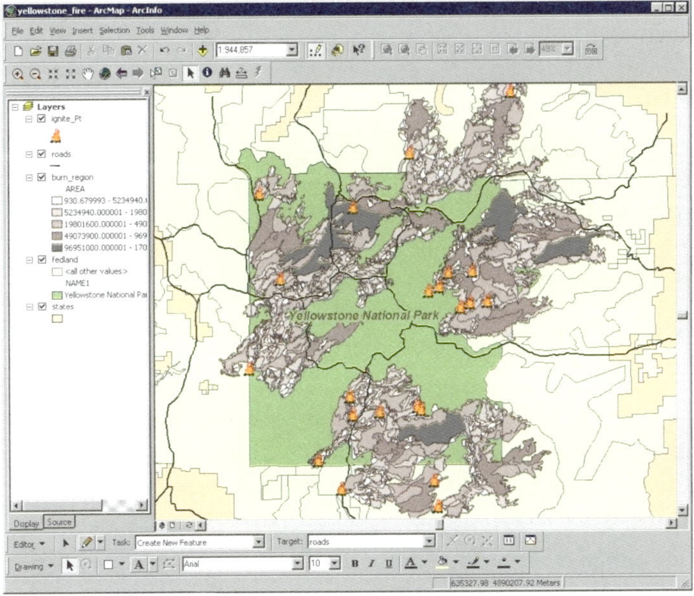

ArcMap is the application in ArcGIS Desktop used to perform most of the map-based tasks including cartography, data analysis, and editing.

Presentation of geocoded addresses

ArcMap is where you will visually interpret and present your geocoding results. A variety of tools allow you to manipulate symbology, scale dependence, and classification based on specific characteristics of each point location.

Geocoding a single address interactively with ArcMap

You can also perform your geocoding process within ArcMap. Geocoding both tables and individual addresses in ArcMap provides a method of instantly presenting the address location on your map.

Interactive review of geocoded addresses

When reviewing your geocoded addresses, ArcMap provides you with the tools to assess the accuracy of the points created.

Creating maps showing analysis of geocoded data

ArcMap provides a suite of cartographic tools to create maps. You can add your geocoded addresses, analysis results, and other related geographic data to a printable map layout. You can then add other map features such as a title, North arrow, neatline, legend, scalebar, and so on.

Toolboxes

Toolboxes store many GIS tools used in geoprocessing. These tools and wizards facilitate basic and advanced geoprocessing tasks such as creating buffers, importing and exporting data types, and manipulating data formats and projections. Within the collection of tools, one toolbox is dedicated entirely to the geocoding process to facilitate many common tasks.

Toolboxes can be opened within ArcMap or ArcCatalog. For more information on the geoprocessing functionality available in ArcGIS, see *Geoprocessing in ArcGIS*.

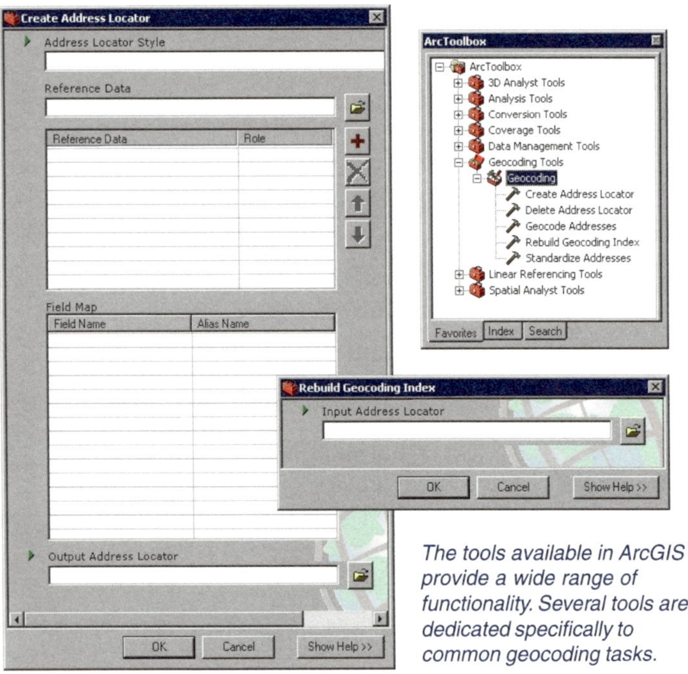

The tools available in ArcGIS provide a wide range of functionality. Several tools are dedicated specifically to common geocoding tasks.

Cleaning reference material

As a valuable tool in the geocoding process, geoprocessing tools include a tool specifically designed to modify your reference material to work harmoniously with the address locator.

Using geoprocessing tasks within a model

The geoprocessing functionality provides a user interface in which a series of tools can be combined into a model-like structure. Geocoding tools can play an integral part within these models.

Analyzing geocoded addresses

A large suite of tools are also available that can assist you in the analysis of your geocoded addresses. These include everything from creating a buffer around the address locations to interpreting clusters.

ArcSDE

ArcSDE is the key component in a multiuser ArcGIS system. It allows you to manage geographic information and distribute it to ArcGIS Desktop, ArcIMS®, and other applications. As part of the data management functionality, you can create address locators and share them across these multiple applications.

ArcGIS Server

While ArcSDE was designed to share geographic information through a local network, ArcGIS Server provides a mechanism in which such data can be shared across a wide range of mediums. These include Web applications, shared networks, Java applications, and so forth.

For a more complete description of each of these applications and further discussion on their interactions, see *What is GIS?*

Tips on learning the geocoding process

If you're new to GIS, take some time to familiarize yourself with ArcMap and ArcCatalog. The books *Using ArcMap* and *Using ArcCatalog* contain tutorials to show you how to make maps and manage GIS data.

Begin learning the process of geocoding in Chapter 2, 'Quick-start tutorial'. In Chapter 2 you'll learn some of the basic and more complex processes of geocoding. The data required for the tutorial is provided with the ArcGIS Desktop software, so you can follow along step by step at your computer. You can also read the tutorial without using your computer.

Finding answers to questions

Like most people, your goal is to complete your task while investing a minimum amount of time and effort in learning how to use the software. You want intuitive, easy-to-use software that gives you immediate results without having to read pages and pages of documentation. However, when you do have a question, you want the answer quickly so that you can complete your task. That's what this book is all about—getting you the answers you need, when you need them.

This book describes many of the concepts of geocoding, how to build and use address locators, and methods of sharing the geocoding functionality. Although you can read this book from start to finish, you'll likely use it more as a reference. When you want to know how to do a particular task, such as creating an address locator, just look it up in the table of contents or the index. You'll find a concise, step-by-step description of how to complete the task. Some chapters also include detailed information that you can read if you want to learn more about the concepts behind the tasks. You can also refer to the glossary in this book if you come across any unfamiliar terms.

Getting help on your computer

In addition to this book, use the ArcGIS Desktop Help system to learn the process of preparing for and performing geocoding. To learn how to use Help, see *Using ArcMap*.

Contacting ESRI

If you need to contact ESRI for technical support, refer to 'Contacting Technical Support' in the 'Getting more help' section of the ArcGIS Desktop Help system. You can also visit ESRI on the Web at *www.esri.com* and *support.esri.com* for more information on geocoding and ArcGIS.

ESRI education solutions

ESRI provides educational opportunities related to geographic information science, GIS applications, and technology. You can choose among instructor-led courses, Web-based courses, and self-study workbooks to find educational solutions that fit your learning style. For more information go to *www.esri.com/education*.

Quick-start tutorial

2

IN THIS CHAPTER

- **Exercise 1: Locating and rematching addresses**
- **Exercise 2: Using alternate street names and place name aliases**
- **Exercise 3: Creating a dynamic geocoded feature class**

This tutorial introduces you to geocoding with ArcGIS Desktop. You will learn the basic techniques for creating address locators and geocoding with ArcGIS. Exercises 2 and 3 demonstrate some advanced features—using alternate names and place name aliases and creating a dynamic feature class that is related to the address table.

Some procedures in this tutorial require that you be familiar with using ArcCatalog and ArcMap. For example, you should know how to copy data to a new location and add data to a map. You should also have a basic understanding of what a geodatabase is and the objects it can contain. If you're new to GIS or feel you need to refresh your knowledge, please take some time to read *Getting Started with ArcGIS*. You might also find that working through the quick-start tutorials in *Using ArcCatalog* and *Using ArcMap* will help you understand many of these basic concepts.

Exercise 1: Locating and rematching addresses

In the first part of this tutorial you'll learn how to create a new address locator, how to locate an address on a map, and how to create point features representing street addresses stored in a table such as customer locations.

Before you begin, it is recommended that you make a copy of the data used in this tutorial so the original tutorial data will remain unmodified. Copy the Atlanta folder to a new location on your computer such as the C:\ disk; you'll find it in the ArcTutor\Geocoding folder on the local disk where the tutorial data was installed. You'll need 5 MB of free disk space to store a copy of the Atlanta folder.

The Atlanta folder contains a personal geodatabase named Atlanta.

Creating a new address locator

An address locator lets you convert textual descriptions of locations into geographic features. The Address Locators folder near the top of the Catalog tree lets you manage existing locators and create new address locators on your computer. The first thing you'll do is create an address locator based on your copy of the Atlanta personal geodatabase.

1. Start ArcCatalog.
2. In the Address Locators folder near the top of the Catalog tree, double-click Create New Address Locator.

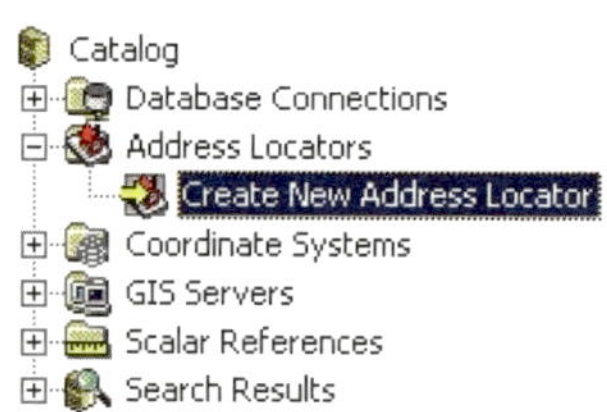

The Create New Address Locator dialog box appears.

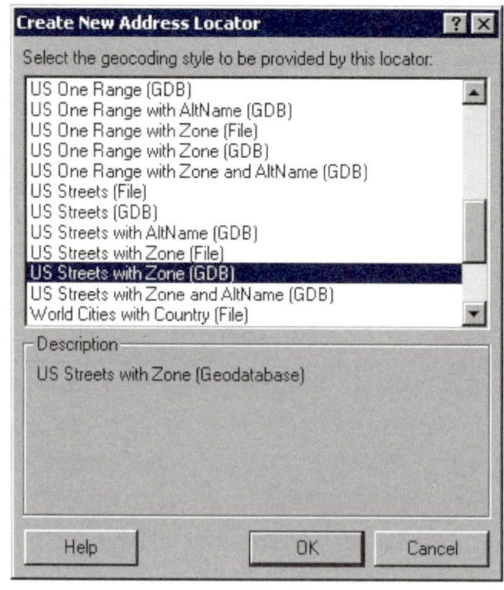

3. Click the US Streets with Zone (GDB) geocoding style, then click OK.

   The New US Streets with Zone (GDB) Address Locator dialog box appears.

4. In the Name text box, type “Atlanta”, replacing the default name.

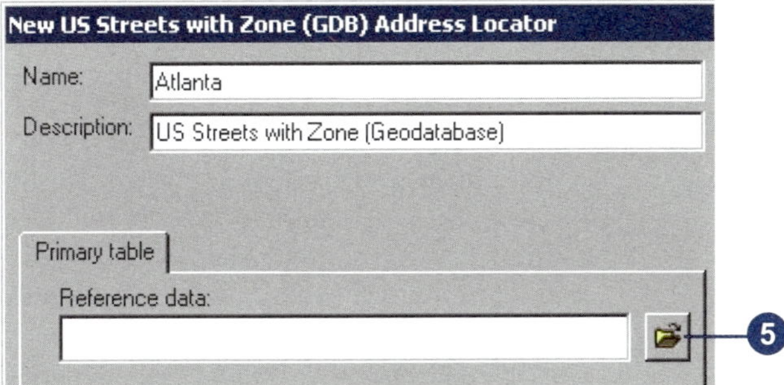

5. On the Primary table tab, click the Browse button next to the Reference data text box. This will open the Choose Reference Data dialog box.

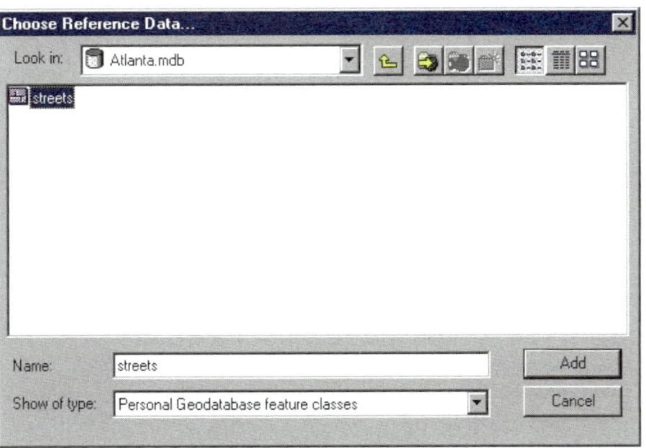

6. In the Choose Reference Data dialog box, navigate to the folder containing your copy of the Atlanta personal geodatabase. Double-click the personal geodatabase, click the streets feature class, then click Add. This will add the streets feature class as reference data to the Primary table tab and populate the fields.

7. Leave all other address locator settings as they are and click OK in the New US Streets with Zone (GDB) Address Locator dialog box. The new address locator appears in the Address Locators folder.

   You’ll notice that the name of the address locator is prefaced by your login name. If your login name is “ssmith”, the address locator would appear in the Catalog tree as “ssmith.Atlanta”.

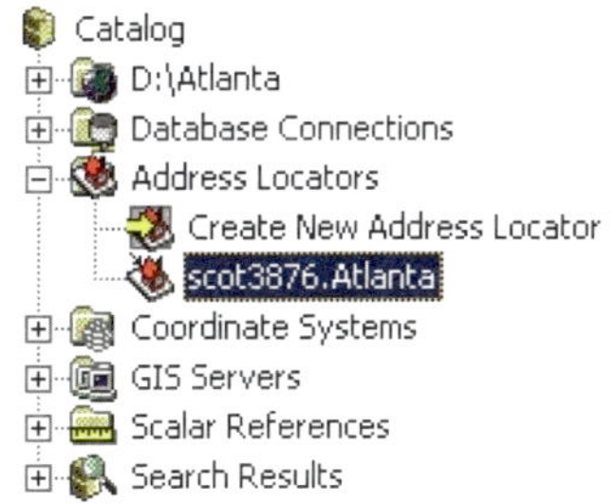

New address locators are private; other users who log in to the same computer won’t have access to your address locators.

When the new address locator is created, a geocoding index is built for the streets feature class within the Atlanta geodatabase. The index is stored in a new table named GC_SZS1 in the Atlanta geodatabase. With this index, the street features can be quickly matched to the addresses you want to geocode.

You will now modify the properties of the address locator you just created.

8. Right-click the yourLogin.Atlanta address locator and click Properties. The Address Locator Properties dialog box appears.

   On the right under Matching Options, you can see that the default Minimum candidate score is 10. This setting determines the matching score needed to return a potential candidate.

9. Click and drag the slide bar to change the Minimum candidate score to a value of 30.

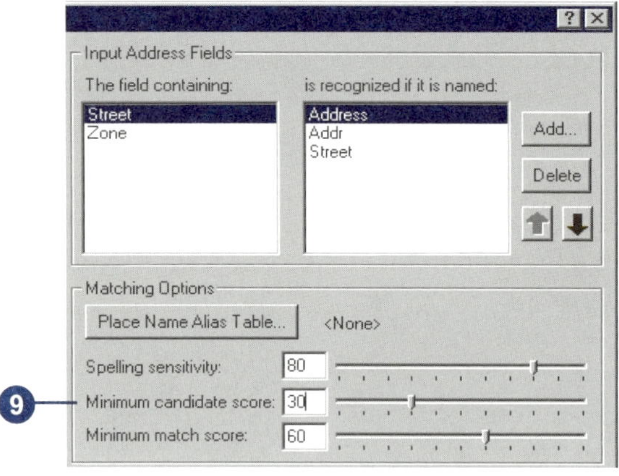

10. Click OK, closing the Address Locator Properties dialog box.

Your changes are saved in the address locator. Now you can use this address locator to find out where addresses are located on a map.

Finding an address interactively

You can use address locators to locate addresses while you are working in ArcMap. This process is commonly called geocoding.

1. Start ArcMap and click OK to create a new, empty map.
2. Add the streets feature class in the Atlanta geodatabase to the ArcMap document.

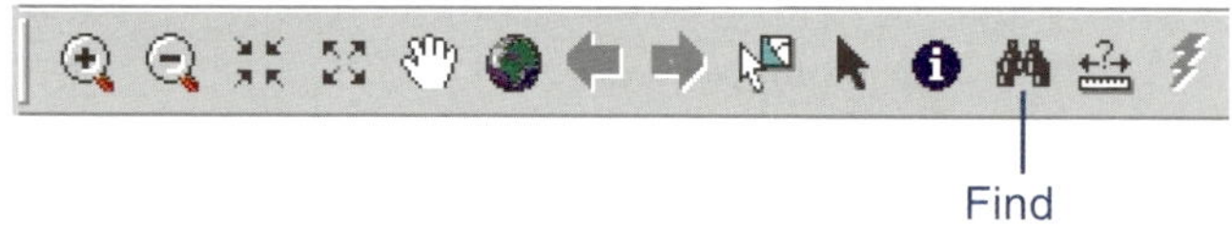

3. Click the Find button on the Tools toolbar.
4. In the Find dialog box, click the Addresses tab.

   Before you can locate an address on your map, you must identify the address locator you want to use.

5. Click the Browse button to choose an address locator. Navigate to the Address locators folder, click the yourLogin.Atlanta address locator, and click Add.

   This populates the Choose an address locator text box and adds two additional text boxes that will be used to enter the address elements.

6. Type "150 Linden Ave NE" in the Street or Intersection text box. Type "30308" in the Zone text box.

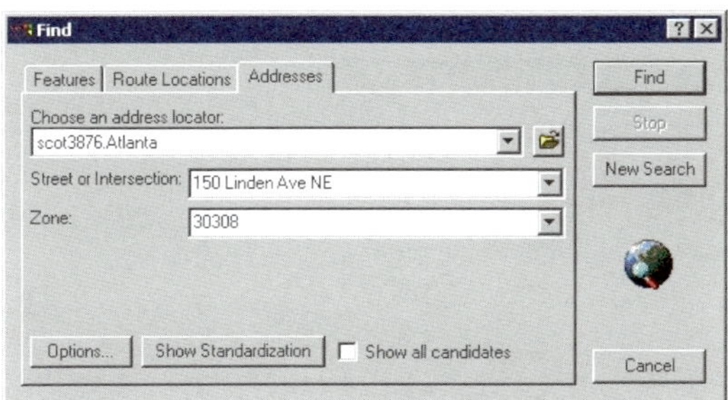

7. Click Find.

   One candidate address appears in the list at the bottom of the Find dialog box.

8. Right-click the candidate and click Add as Graphic(s) to Map.

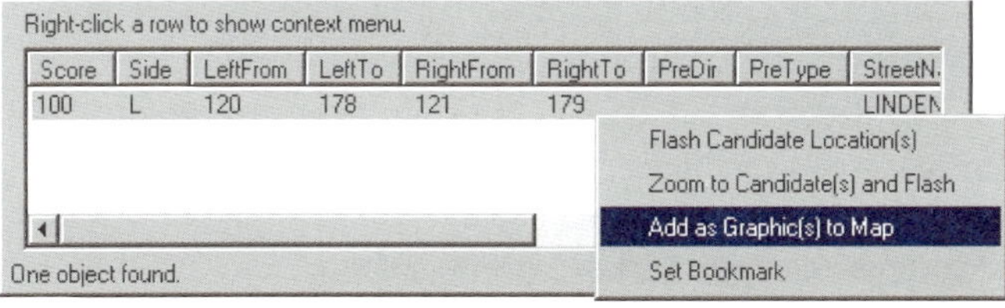

A graphic point representing the address's location appears on the map.

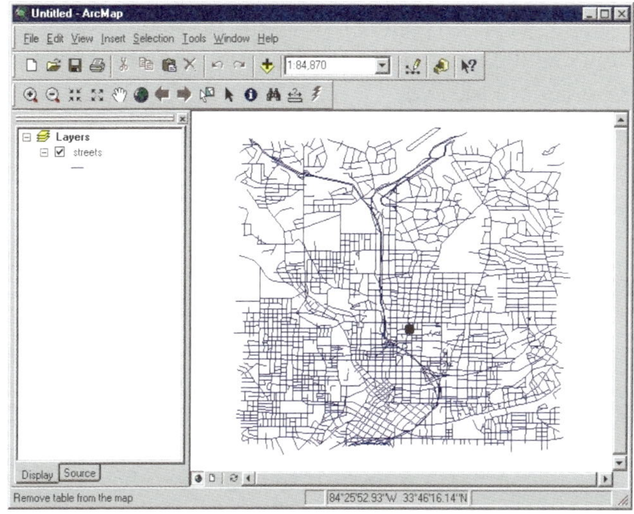

9. In the Find dialog box, check Show all candidates.

10. Click the Close button in the upper right corner of the Find dialog box.

Geocoding addresses in a table

If you have a table that contains address information, you can geocode all of the addresses at once. In this exercise you will begin with a table containing the addresses of customers and geocode the entire table, showing where the customers are located.

1. Open a new map document in ArcMap. You do not need to save the map you created in the previous task.
2. Add the streets feature class and the customers table from your copy of the Atlanta geodatabase to your map.
3. Right-click the customers table on the Source tab of the table of contents and click Geocode Addresses.
4. On the Choose an address locator dialog box, click Add.

   The Add Address Locator dialog box will open.
5. Navigate to the Address Locators folder, click the yourLogin.Atlanta address locator, then click Add.
6. Click yourLogin.Atlanta in the Choose an address locator to use dialog box, then click OK.

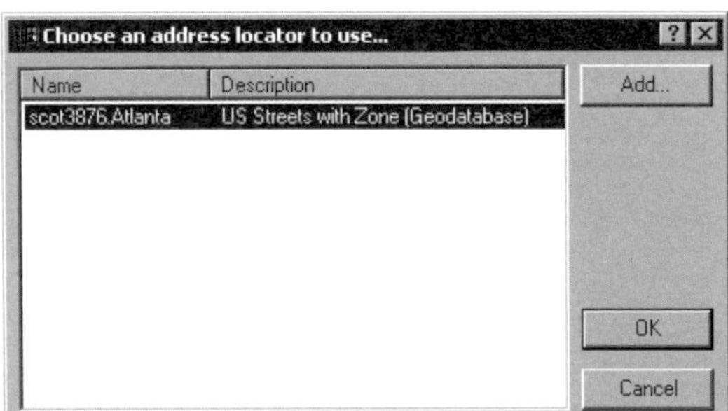

   This will open the Geocode Addresses: yourLogin.Atlanta dialog box.

7. Under the Output section, click the Browse button to define the Output shapefile or feature class. This will open the Saving Data dialog box.

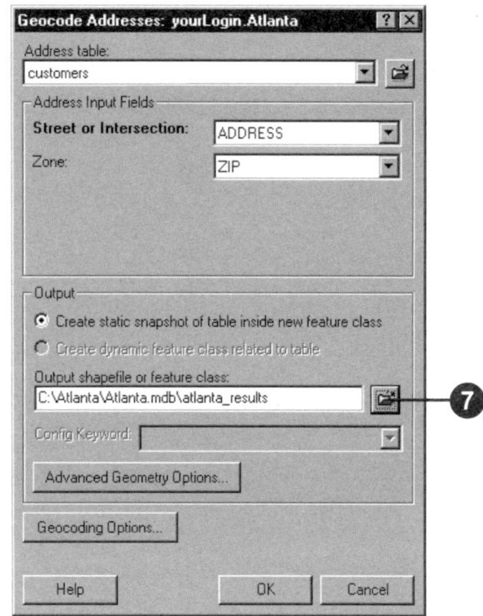

8. Click the Save as type dropdown arrow, then click Personal Geodatabase feature classes.

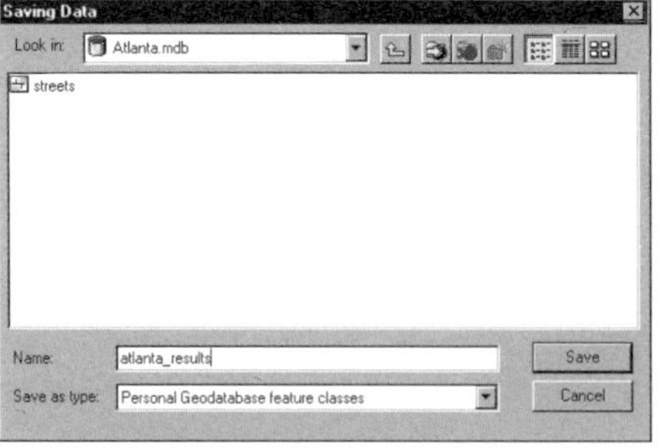

9. Navigate to your copy of the Atlanta personal geodatabase, then double click the personal geodatabase. In the Name text box, type "atlanta_results".
10. Click Save.

    A new point feature class named atlanta_results will be created in the Atlanta geodatabase. The point features generated by the geocoding process will be saved in that feature class.
11. Click OK to start geocoding.

    When the geocoding process is finished, the Review/ Rematch Addresses dialog box appears. It shows the statistics of the addresses that were matched or not matched. You'll rematch the unmatched addresses later in the tutorial.
12. Leave the rematch criteria as is and click Done to close the Review/Rematch Addresses dialog box.

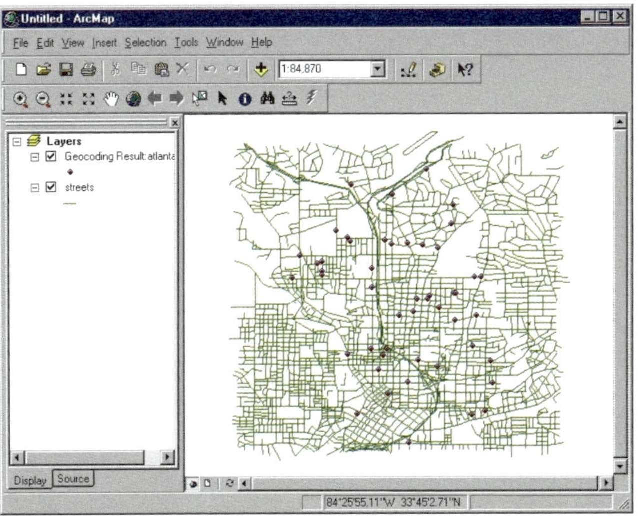

A Geocoding Result layer is added to the map. It shows the points that were added to the atlanta_results feature class.

You can also geocode a table of addresses in ArcCatalog. To do so, right-click the customers table in the Atlanta geodatabase and click Geocode Addresses. This will induce a series of dialog boxes identical to those identified in this exercise. You can then add the created feature class to an ArcMap document.

Rematching addresses

1. In ArcMap, click the Tools menu, point to Geocoding, point to Review/Rematch Addresses, then click Geocoding Result: atlanta_results.
2. The message "This operation requires that you be editing this workspace. Would you like to start editing?" appears. Click Yes.

   The Review/Rematch Addresses dialog box opens.
3. In the Rematch Criteria section, click the option Addresses with candidates tied.

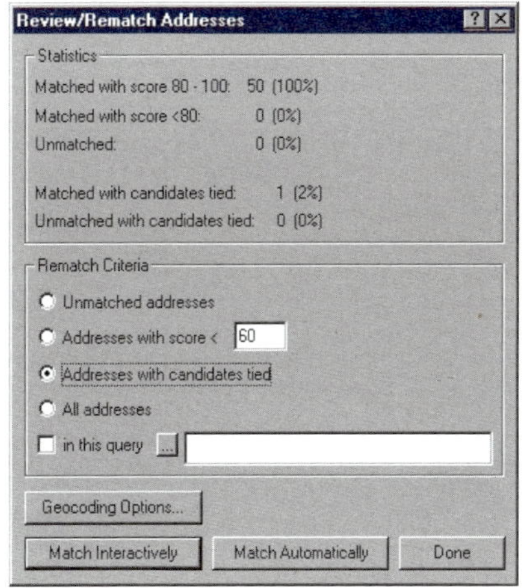

4. Click Match Interactively.

   The Interactive Review dialog box appears. The address records that you have chosen to rematch are listed at the top of the dialog box. In this exercise there is one record in the feature class that has two candidates that are tied for the highest match score. The tied candidates are listed at the bottom of the dialog box.

5. Arrange the ArcMap window and the Interactive Review dialog box so you can see the map.

   Near the center of the map, you will notice two points highlighted, one in yellow and the other in blue. These are the locations of the two candidates listed in the bottom window of the Interactive Review dialog box. The selected candidate is highlighted in yellow on the map.

6. Click the second candidate in the candidate list. The location of the yellow highlighted candidate changes on the map.

7. In the bottom right corner of the Interactive Review dialog box, click Match. The selected address in the list at the top of the dialog box is now associated with the second candidate in the list at the bottom.

8. Click the Close button to close the Interactive Review dialog box.

9. Click Done to close the Review/Rematch Addresses dialog box.

10. On the Editor toolbar, click the Editor menu and click Stop Editing. Click Yes to save your changes.

11. Close ArcMap. You do not need to save the changes to your map document.

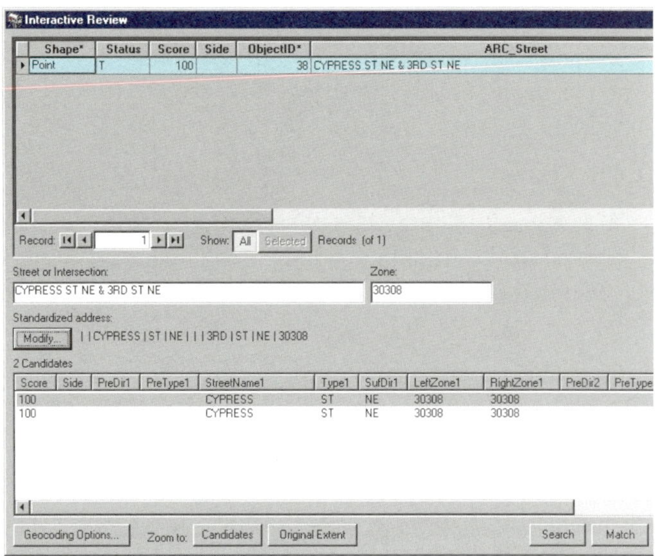

You can also rematch addresses with ArcCatalog. To do so, refresh the contents of the Atlanta geodatabase if you don't see the atlanta_results feature class listed in the Contents tab. Right-click the atlanta_results feature class, then click Review/Rematch Addresses. This will induce a series of dialog boxes similar to those described in this exercise.

Exercise 2: Using alternate street names and place name aliases

When you create an address locator, you have the option to use alternate street names and place name aliases. Place name aliases let you associate names of well-known places with street addresses—for example, the names of museums, hospitals, or landmarks. When geocoding an address, the place name alias table is consulted first. If a place name is found, the street address from the place name alias table is geocoded. Similarly, alternate street names can be defined for features in the reference data. When geocoding an address, the alternate street name table is also searched to find potential candidates where streets have more than one name.

Creating an address locator that uses alternate names and place name aliases

In this task you'll create a new address locator that uses alternate street names and place name aliases, then use it in ArcMap to find addresses.

1. In ArcCatalog, click the Address Locators folder at the top in the Catalog tree, then double-click Create New Address Locator.
2. Click the US Streets with AltName (GDB) geocoding style, then click OK.

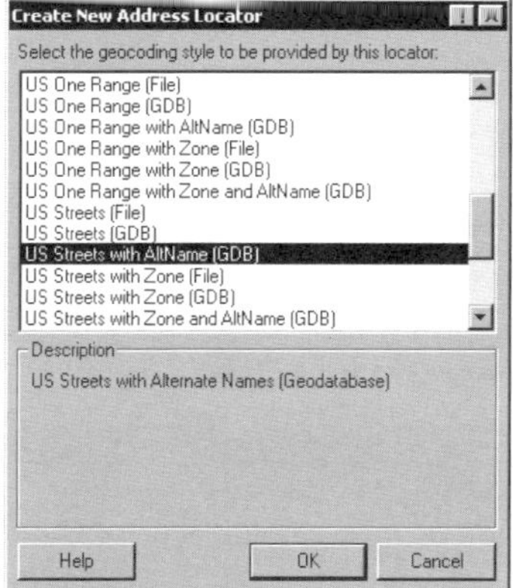

The New US Streets with AltName (GDB) Address Locator dialog box appears.

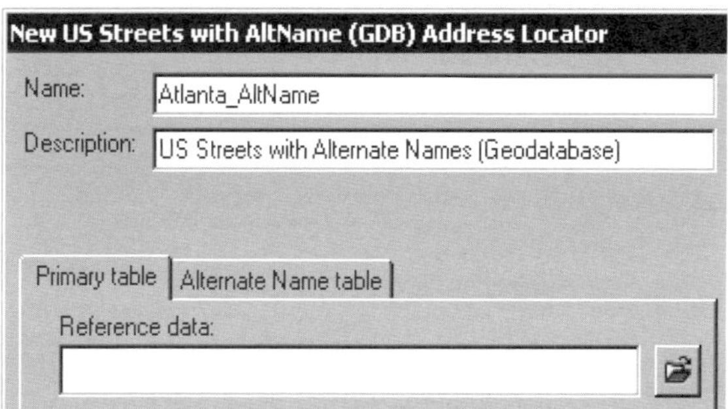

3. In the Name text box, type "Atlanta_AltName", replacing the default name.
4. On the Primary table tab, click the Browse button next to the Reference data text box.

5. From the Choose Reference Data dialog box, navigate to the Atlanta personal geodatabase. Double-click the personal geodatabase, click the streets feature class, then click Add.
6. Click the Alternate Name table tab.
7. Click the Browse button next to the Reference data text box.
8. In the Choose Reference Data dialog box, navigate to your copy of the Atlanta personal geodatabase. Double-click the personal geodatabase, click the altname table, then click Add.
9. On the Alternate Name table tab, click the Join ID dropdown list, then click ALTNAME_ID.

   The ALTNAME_ID column in the altname table and the OBJECTID column in the streets feature class are used to relate the records in the altname table to the records in the streets feature class.

10. On the right under Matching Options, click Place Name Alias Table.

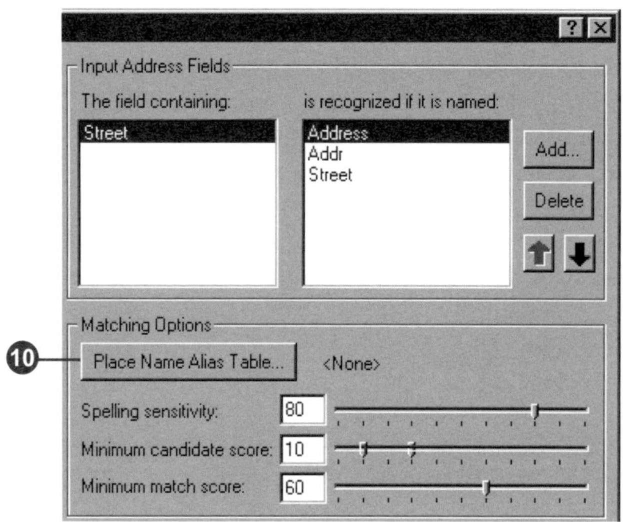

   The Alias Table dialog box appears.

11. Click the Browse button, then navigate to the Atlanta geodatabase. Double-click the personal geodatabase, click the place_aliases table in the Atlanta geodatabase, then click Add.
12. Click the Alias field dropdown arrow, then click NAME.

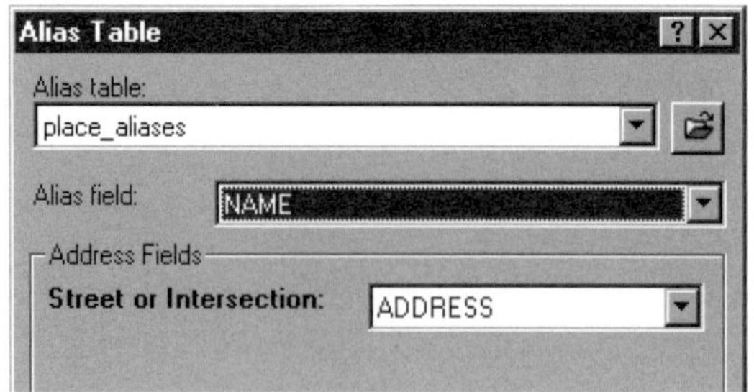

13. Click OK.
14. Click OK in the New US Streets with AltName (GDB) Address Locator dialog box. The new yourLogin.Atlanta_AltName address locator appears in the Address Locators folder.

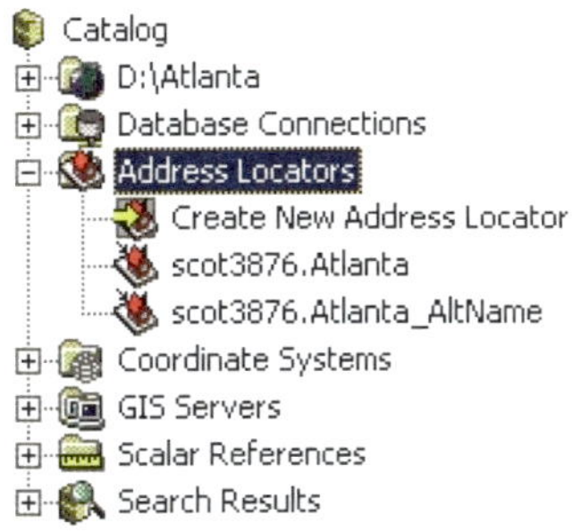

An additional geocoding index table has been added to your copy of the Atlanta personal geodatabase. This new table facilitates searching in the alternate name table.

15. Close ArcCatalog.

You can now use this address locator to find out where an address is located on a map.

Finding locations using alternate street names and place name aliases

1. Open a new map document in ArcMap.
2. Add the streets feature class from the Atlanta geodatabase to your map.
3. Click the Find button on the Tools toolbar.
4. In the Find dialog box, click the Addresses tab.
5. Click the Browse button to choose an address locator. Navigate to the Address Locators folder, click the yourLogin.Atlanta_AltName address locator, and click Add.

   A street may have more than one name. For example, “Atlanta Blvd” is defined as an alternate name for “Old 10th St NE” in the Atlanta database. With the yourLogin.Atlanta_AltName address locator, “30 Old 10th St NE” and “30 Atlanta Blvd” will both find the same location.
6. Type “30 Old 10th St NE” in the Street or Intersection text box, then click Find.

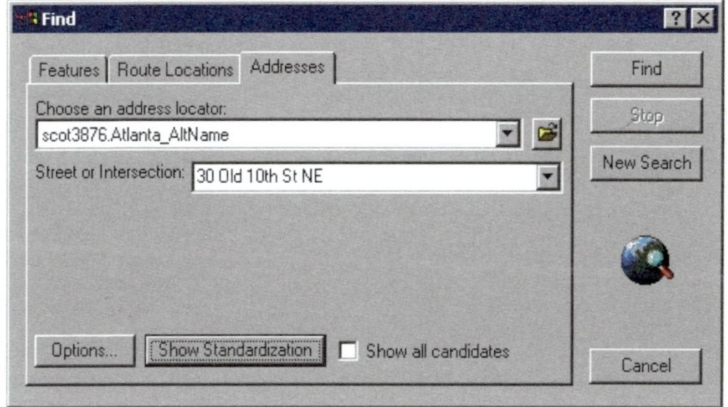

7. Right-click the candidate and click Add as Graphic(s) to Map.

   A graphic point representing the address’s location appears on the map.
8. Type “30 Atlanta Blvd” in the Street or Intersection text box, then click Find.

9. Right-click the candidate and click Flash Candidate Location(s). You can see that both addresses are matched to the same location on the map.

   In addition to geocoding addresses, you can also geocode place names that have been defined in the place_aliases table.

10. Type “Children’s Hospital” in the Street or Intersection text box, then click Find.

11. Right-click the first candidate in the candidates list, which has a score of 100, and click Add as Graphic(s) to Map.

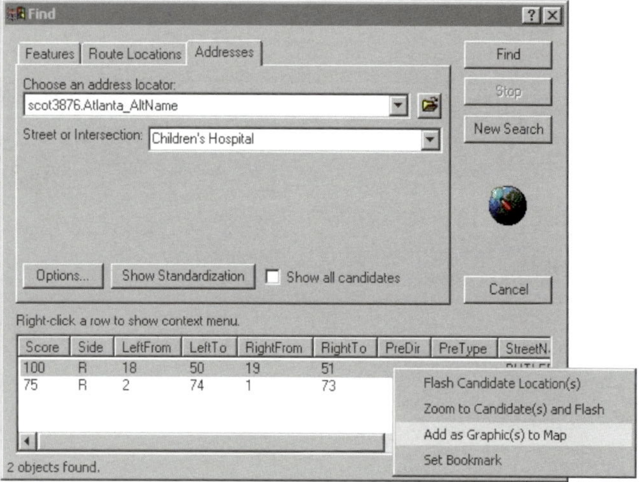

The location of the hospital appears on the map as if you had typed its complete address.

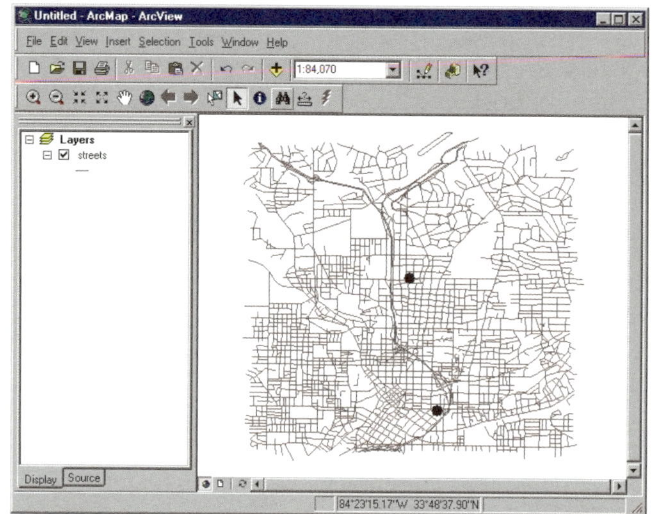

12. Click Cancel to close the Find dialog box.

Creating address locators that use alternate names and place name aliases is beneficial for people who know a landmark’s name but not its address. It will also help in situations where a street has more than one name.

Exercise 3: Creating a dynamic geocoded feature class

With an ArcInfo™ or ArcEditor™ seat, your geocoding results can be dynamically linked to the original addresses if the address table and the result feature class are stored in the same geodatabase. With dynamic results, modifications to the primary address table will cause the geocoded feature class to be updated automatically. ArcView® users will not be able to complete this exercise.

1. Open a new map document in ArcMap. You do not need to save your previous map document.
2. Add the customers table and the streets feature class in the Atlanta geodatabase to your map.
3. Right-click the customers table in the Source tab of the map's table of contents and click Geocode Addresses.
4. Click Add in the Choose an address locator to use dialog box.
5. In the Add Address Locator dialog box, navigate to the Address Locators folder, click the yourLogin.Atlanta address locator, then click Add.
6. Click yourLogin.Atlanta in the Choose an address locator to use dialog box, then click OK.

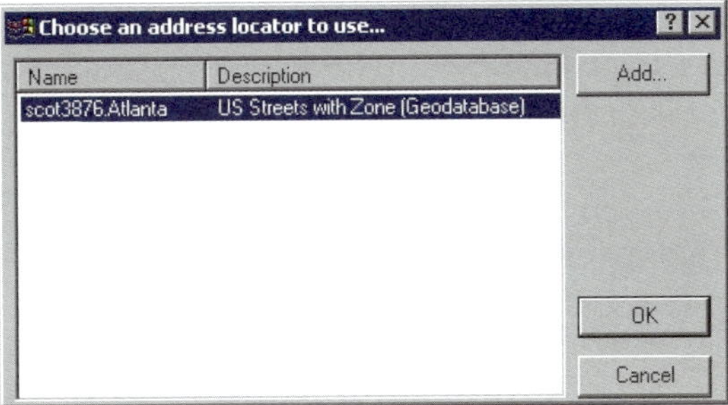

7. In the Geocode Addresses: yourLogin.Atlanta dialog box, click the Browse button to define the Output feature class.
8. In the Saving Data dialog box, click the Save as type dropdown arrow and click Personal Geodatabase feature classes.
9. Navigate to the Atlanta personal geodatabase, then double-click the personal geodatabase.
10. In the Name text box, type "dynamic_results". Click Save.

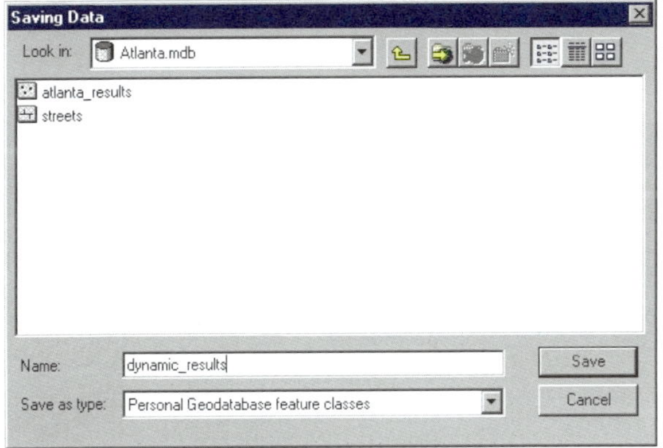

11. Click Create dynamic feature class related to table.

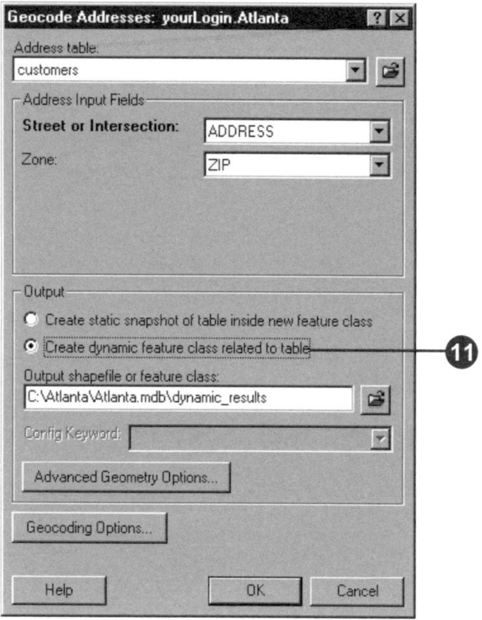

A new point feature class named dynamic_results will be created in the Atlanta geodatabase. The point features generated by the geocoding process will be saved in that feature class. A relationship class between the table and feature class will be created.

12. Click OK to start geocoding.

When the geocoding process is finished, the Review/ Rematch Addresses dialog box appears.

13. Click Done to close the Review/Rematch Address dialog box.

A Geocoding Result layer is added to the map. It shows the points that were added to the dynamic_results feature class.

14. Right-click the Geocoding Result layer in the map's table of contents and click Open Attribute Table.

Attributes of Geocoding Result: dynamic_results

Shape*	Status	Score	Side	ObjectID*	Street or Intersection
Point	M	100	R	1	1171 PIEDMONT AVE NE
Point	M	100	L	2	1670 W PEACHTREE ST NE
Point	M	100	R	3	455 BEVERLY RD NE
Point	M	100	R	4	241 16TH ST NW

The first record in the table has the address 1171 PIEDMONT AVE NE in the Street or Intersection column. The Status (M: matched, U: unmatched, or T: tied), Score (match score), and Side (R: right side of the road or L: left side of the road) columns contain the values M, 100, and R respectively. To see where this customer is located on the map, click this record in the table. The feature will be highlighted on the map.

15. Minimize the Geocoding Result layer's attribute table.

16. Right-click the customers table in the map's table of contents and click Open.

The address that appears for the first customer in the table, Ace Market, is 1171 Piedmont Ave. NE. That customer has moved to a new address, 30 12th St. The ZIP Code remains the same. The customers table must be updated with this information.

17. If the Editor toolbar is not showing, click the Editor Toolbar button to show the Editor toolbar.

18. On the Editor toolbar, click the Editor menu and click Start Editing.

19. Type "30 12TH ST" in the Address column next to Ace Market.

Attributes of customers

OBJECTID*	NAME	ADDRESS
1	Ace Market	30 12TH ST
2	Andrew's Gasoline	1670 W PEACHTREE ST NE
3	AP Supermarket	455 BEVERLY RD NE
4	Atlanta Market	241 16TH ST NW

20. Click another record in the table.
21. Minimize the customers table and maximize the Geocoding Result layer's attribute table.

The new address was geocoded automatically, and the corresponding values have been updated in the layer's attribute table. The Status, Score, and Side columns now contain the values T, 81, and L, respectively. The value "T" in the Status field stands for tied; two or more candidates had the highest score. To see on the map where Ace Market is now located, click this record in the layer's attribute table.

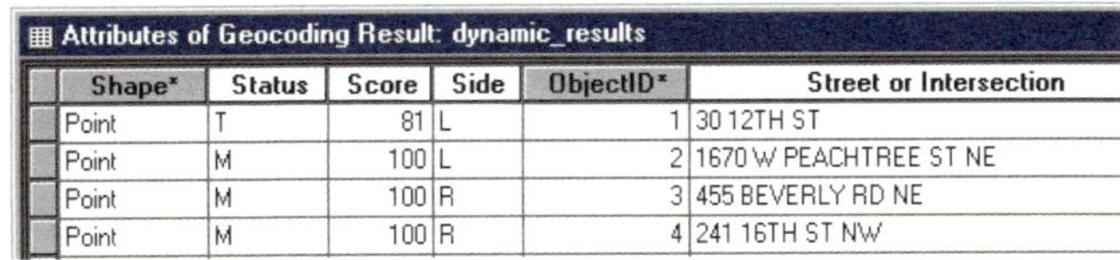
Attributes of Geocoding Result: dynamic_results

Shape*	Status	Score	Side	ObjectID*	Street or Intersection
Point	T	81	L	1	30 12TH ST
Point	M	100	L	2	1670 W PEACHTREE ST NE
Point	M	100	R	3	455 BEVERLY RD NE
Point	M	100	R	4	241 16TH ST NW

Now you need to add a new customer to the customers table.

22. Minimize the Geocoding Result layer's attribute table and maximize the customers table.
23. Scroll down to the bottom of the customers table.
24. In the blank record after the last record in the table, click in the NAME column. Type "Vine Cafe".

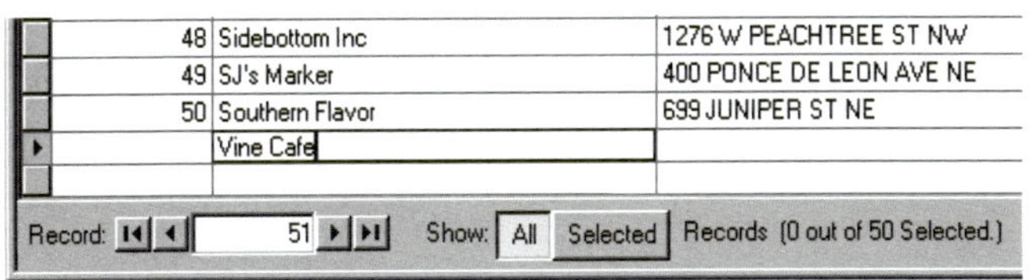

48	Sidebottom Inc	1276 W PEACHTREE ST NW
49	SJ's Marker	400 PONCE DE LEON AVE NE
50	Southern Flavor	699 JUNIPER ST NE
	Vine Cafe	

Record: 51 Show: All Selected Records (0 out of 50 Selected.)

A new blank record is added to the bottom of the table.

25. In the same record, type "379 VINE ST NW" in the ADDRESS column.
26. Type "30314" in the ZIP column.
27. Type "Cafe" in the TYPE column.
28. In the SALES column type "10000".

29. Close the customers table. Maximize the Geocoding Result layer's attribute table, then scroll to the bottom.

    A new feature was automatically added to the result feature class representing the new address.

30. On the Editor toolbar, click the Editor menu and click Stop Editing. Click Yes to save your changes.

31. Close ArcMap. Click No to stop ArcMap without saving this map.

3 Concepts of geocoding

IN THIS CHAPTER

- **What is an address?**
- **The geocoding work flow**
- **The ArcGIS geocoding framework**
- **The process of geocoding**
- **Customizing your address locator**
- **Quick reference guide to common geocoding tasks**

There are a number of basic concepts that will greatly facilitate your geocoding experience. The aim of this chapter is to introduce you to the standard geocoding work flow and to simplify many of the concepts involved in geocoding. An understanding of these concepts will assist you in understanding how and why you get the results that you do. You will be able to assess your results and choose if and how to modify your search parameters. This chapter concludes with a discussion on the reasons and methods of customizing your address locator and other related files.

What is an address?

An address is simply a method used to describe a location. Unlike a coordinate value, an address describes how to reference a location based on existing features in your GIS database. In most cases, this description is relatively easy to understand. For example, if you needed to locate the address "380 New York St., Redlands, CA 92373" with the correct street data, it would not take you long to find the exact location. You might first find California, then find the city of Redlands. You might also use a postal code map and locate the region covered by the corresponding ZIP Code value. You would then locate the street, and finally interpret where and on which side of the 300 block the address is located.

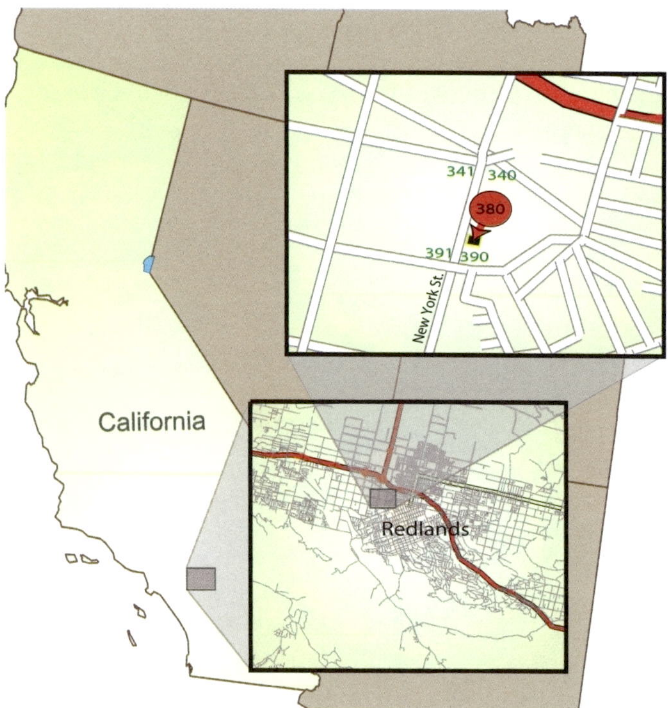

When locating a U.S. address, typically you find the state, then the city, and finally the exact street.

Just as you first narrowed your search to a specific region, found a particular feature, and finally interpreted a point, the computer is doing the same process to assign a location to an address when geocoding.

Address elements

Addresses have some specific characteristics. An address contains certain address elements and is presented in a range of formats. When geocoding, the address format is interpreted, address elements are identified, and these address elements are compared against elements in the reference data.

An address element is an individual component in the address such as the house number, the street name, and the postal code. Address elements help in the geocoding search, pinpointing an address to a particular location.

Address formats

Addresses are represented in a wide range of formats. A common address format used in the United States consists of the following series of address elements: house number; street name; street type; and zone information such as city, state, and ZIP Code.

26376	Alpine	Lane,	Twin Peaks,	CA	92391
House number	Street name	Street type	City	State	Postal code

A common address format in the United States, consisting of the basic address elements

In many areas, addresses are presented in different formats. One example of these alternative formats is the address format used in Queens, New York. In the mid-1920s, the Topographic Bureau of Queens unified the street names and implemented a hyphenated

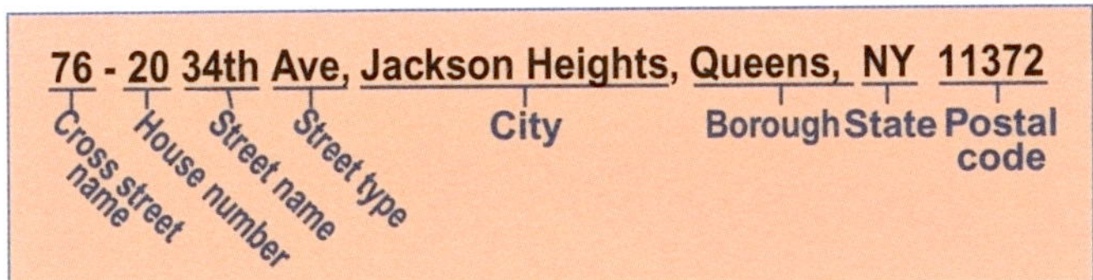

An alternative address format, used in Queens, New York, includes information about the nearest cross street

address style. The first number indicates either the north or west cross street. The second number indicates where on the block the building is located. Also as a general rule in Queens, avenues run east to west and streets run north to south. Queens also includes

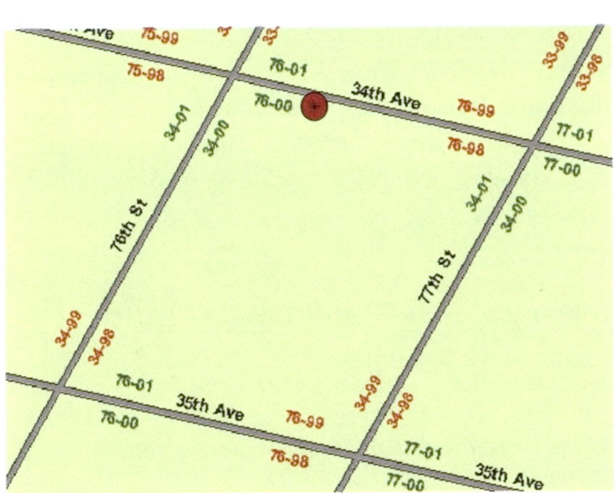

In Queens, New York, the address format contains an additional numeric value representing the nearest cross street.

the neighborhood, or borough, where the address is located. This is a practice used in many parts of the world. While the address format used in Queens is not initially recognizable, the address format still contains the elements needed to assign it to a specific location.

Salt Lake City, Utah, also uses an alternative address format. When the streets were initially laid out in Salt Lake, the Latter-Day Saints temple was the center of the community. Roads in each direction from the temple were assigned a numeric name indicating how far it was from the temple as well as directional value, indicating the direction from the temple that the street was located. The prefix direction simply indicated the part of the road where the address is located. Again, while the address format differs from the common format, the basic address elements exist to locate each address.

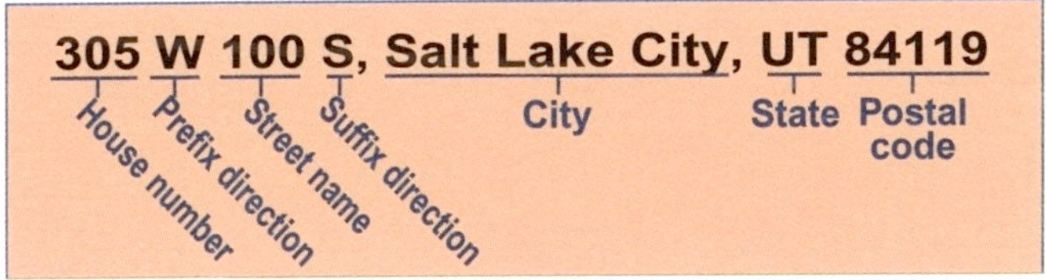

In Salt Lake City, Utah, there are potentially four streets with the name '100.' The particular street is indicated using the suffix direction, stating on which side of the temple the street is located.

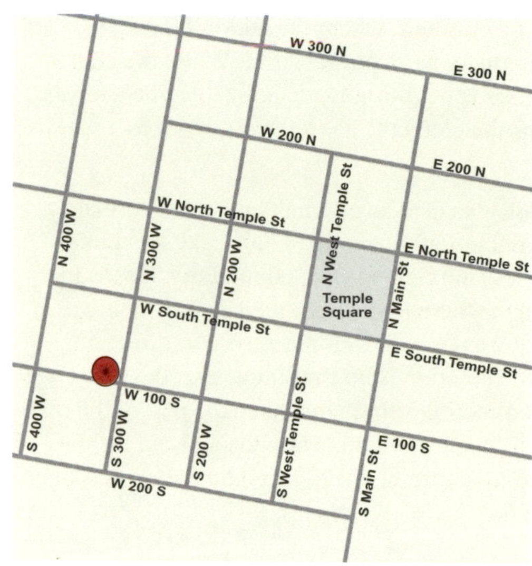

In Salt Lake City, many of the street names are based on the distance from the Latter-Day Saints temple.

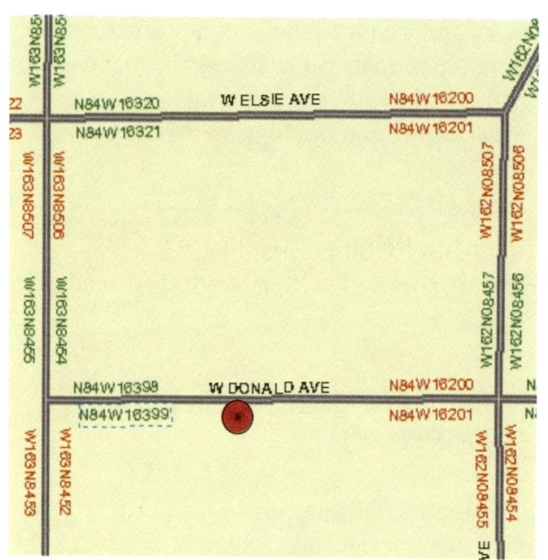

In some regions, a grid code is added to the street address format.

Another United States example of address formats can be found in regions of Illinois and Wisconsin. In these regions, the address format includes a "grid zone" address element. A grid zone is simply a larger block or grid on which the address is located. The corresponding house numbers on the streets are assigned values based on the location and the particular street block. The grid zone value helps to pinpoint the street number to a particular zone within the community.

N84W	16301	W	Donald	Ave,	St Charles,	IL	60175
Grid zone	House number	Street direct.	Street name	Street type	City	State	Postal code

In some communities in Wisconsin and Illinois, the grid zone address element is introduced.

International addresses can also be presented in a range of formats. For example, a common Brazilian address contains most of the basic address elements. However, they are arranged in a somewhat different pattern. Also, at times, the particular state or province is not directly specified. However, it can be derived from the postal code or city.

Rua	Aurora	753,	01209001	Sao Paulo
Street type	Street name	House number	Postal code	City

An address from Brazil contains all of the basic address elements; they are simply arranged differently.

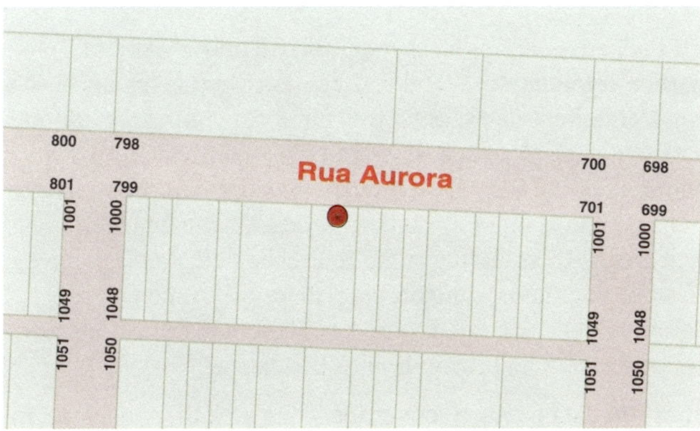

The address format is sometimes rearranged based on the language being used.

Due to some variations in language, it might appear that certain address elements are being eliminated; for example, in German, the street name as well as the street type can be concatenated into one term. Understanding the language and customs of an area will greatly facilitate translation of unfamiliar address formats.

Wendenstrasse 403, 20537 Hamburg

Street name and Street type	House number	Postal code	City
Wendenstrasse	403	20537	Hamburg

In some languages, such as German, address elements, such as street name and street type, can be concatenated together.

While all of these addresses differ to some degree, some things remain consistent. Each address consists of several address elements, presented in a particular address format recognized by those in the region. Understanding that all addresses contain particular address elements will help in understanding the geocoding process and the translation required by each particular geocoding rule base to interpret less common address formats.

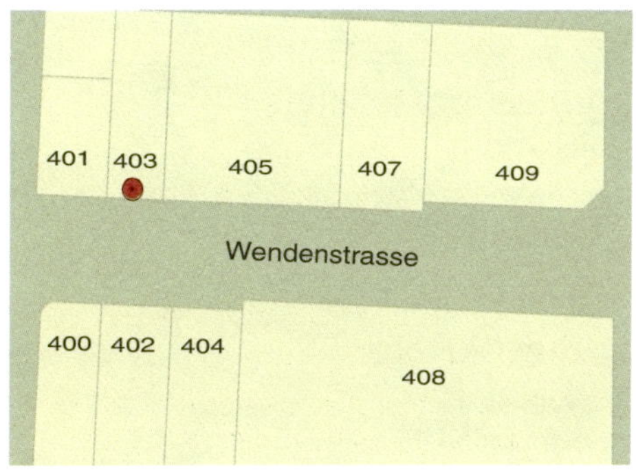

The concatenated term "Wendenstrasse" includes the street name and type.

The geocoding work flow

Building or obtaining reference data

Obtain and make any needed modifications to the reference material to coincide with address locator style requirements.

Determining address locator style

Select an address locator style that accommodates your address data and reference data material as well as produces the desired output information.

Building an address locator

Based on a specific address locator style, create an address locator, incorporating the style-specific guidelines and your reference data.

Locating addresses

Using your address locator, search for an individual address or perform a batch search, locating a group of addresses.

Publishing, maintaining, or customizing your address locator

A wide range of functions are possible including the distribution, updating, and customization of your address locator.

For successful geocoding, there are several steps that should be followed. These steps are referred to as the geocoding work flow. Each of these steps is outlined below.

Building or obtaining reference data

Initially, the geocoding process requires two major types of information, reference data and address data. In this instance, reference data refers to a GIS feature class containing the address attributes you wish to search. For example, when searching for house number addresses, the reference material must contain the street names, house number ranges, or address attributes of the specific parcels. In assessing if the reference data that you have will work for your geocoding process, there are a few considerations that you need to make. These considerations are based on the extent and resolution of the data.

The reference data needs to cover the same area that you want to geocode. For example, if you are attempting to geocode cities across the entire continental United States, reference data that only presents features in the western states will not perform the task. Each address that you want to geocode must be present on your reference data. This leads to the second property of the reference data, the resolution.

Beyond determining if the data has a spatial coverage that includes all of the features you are wanting to geocode, you must also consider if the reference data has information at the detail that you are wanting to search. If you wish to geocode individual addresses, you need to be sure that your reference data has information at this granularity.

While the geometry provides a visual interface on which to compare the results of the geocoding search, it is the set of address attributes in which you will find many of the details that are used in the actual geocoding process. For example, in geocoding specific addresses, the attributes must have information specific to the numeric address values for each feature on the map. Formatting of the attribute table for your reference material will be discussed further in the chapter 'Preparing for geocoding' in this book.

Determining address locator style

The address data refers to the individual or group of addresses you plan to geocode. The address data also needs to comply with certain formats. The specific requirements of the address data are based on the elements of the address locator style that you intend to use. In general terms, the address data needs to contain the desired elements used in matching an address to a feature and be in an acceptable format. Methods for address data presentation are found in the chapter 'Preparing for geocoding' in this book.

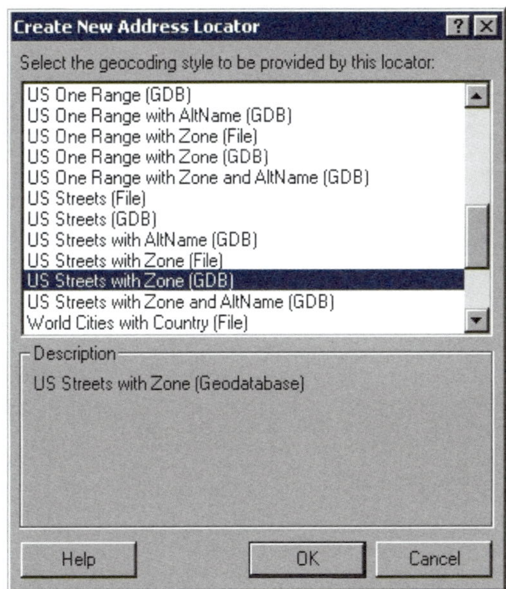

A wide range of geocoding styles are available to use in an address locator. The correct selection of style will be based on the address and reference data formats.

In order to correctly geocode, it is important to understand the variety of address locator styles and to select the one that best matches your address and reference data. The predefined address locator styles are described in the chapter 'Preparing for geocoding' in this book.

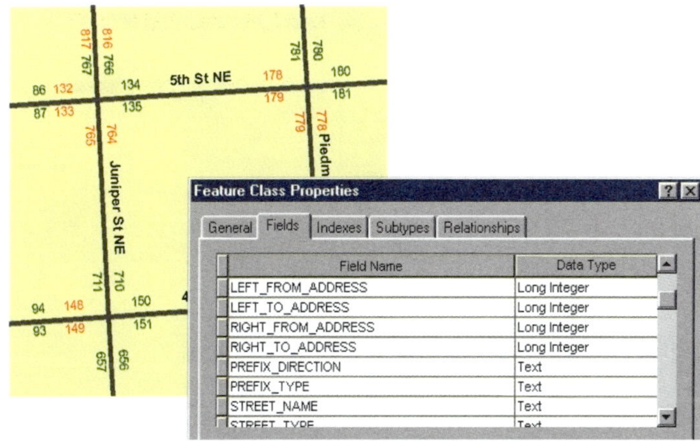

Each geocoding style requires that the primary reference data has certain characteristics, both in its geometry as well as in its attribute table.

Building an address locator

Once an address locator style has been decided and reference and address data have been prepared for that style, you are prepared to build an address locator.

An address locator includes the reference data and guidelines for an address style. These guidelines specify the rules for matching addresses to the reference data as well as suggested adjustments made to enable fuzzy address matching. More information about these guidelines is provided in the chapter 'Modifying your address locator' in this book.

The reference data used in the geocoding process consists of the features with associated address attributes as well as geocoding indexes and geocoding-specific tables used to ensure high performance. For more information on geocoding indexes, see the chapter 'Keeping an address locator current' in this book.

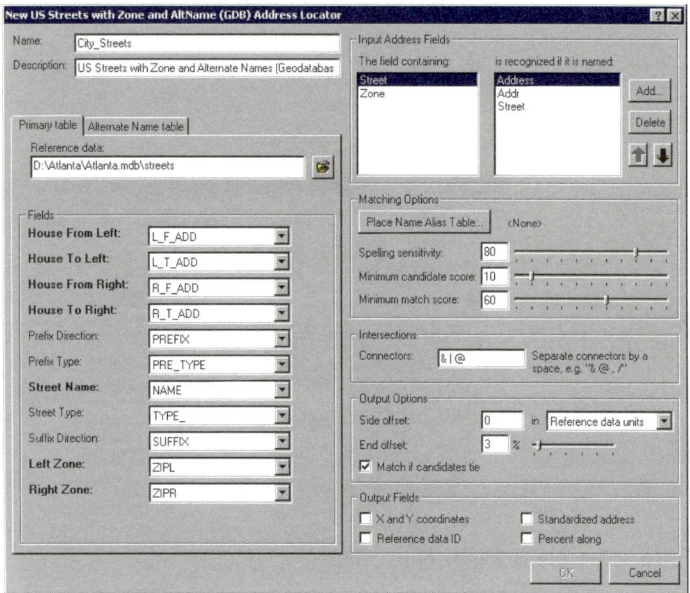

The Address Locator dialog box provides an interface where many adjustments and settings can be made.

Locating addresses

ArcGIS provides two methods for locating addresses. You can find individual addresses or find locations for a table of addresses, commonly referred to as "batch geocoding".

There are several basic steps to locating addresses. These are outlined in the section 'The process of geocoding' in this chapter. In essence, the process includes finding candidate locations in the reference data, assigning match scores to the locations, and narrowing down the best candidate.

Publishing, maintaining, or customizing your address locator

Once you have created an address locator, there are several optional tasks that you may do. These include sharing or publishing the address locator, maintaining the reference data and related address locator, and customizing the user interface or functionality of the address locator.

Web services are one method that can be used to publish the functionality of an address locator.

Publishing the address locator

One common task is the distribution of an address locator. There is a wide range of methods to distribute an address locator. These include sharing the material over a local network, reproducing an address locator and the related files onto a compact disc or into a compressed data file, and producing a geocoding Web service.

Information on distribution techniques is included in the chapter 'Distributing your address locator' in this book.

Maintaining the address locator

When you initially create your address locator, the geocoding indexes reflect the status of your reference data at the time of creation. However, like most data files, your reference data may require frequent updates.

The addition of streets to the data collection, changing street names or zone properties, or simply improving the accuracy of the geometry of the features all justify modification to the geocoding reference data. When these changes are made to the reference data, the geocoding indexes also need to be updated to reflect these changes. Details on maintaining the geocoding indexes are provided in the chapter 'Keeping an address locator current' in this book.

Customizing elements in the address locator

There are a number of methods that can be used to modify your search. These are discussed in the section 'Customizing your address locator' later in this chapter. The degree to which you alter your search ranges from basic changes in the address locator dialog box to redefining your rule base.

The ArcGIS geocoding framework

Understanding the geocoding framework will further help in understanding what the address locator is as well as its interaction and function in the geocoding process.

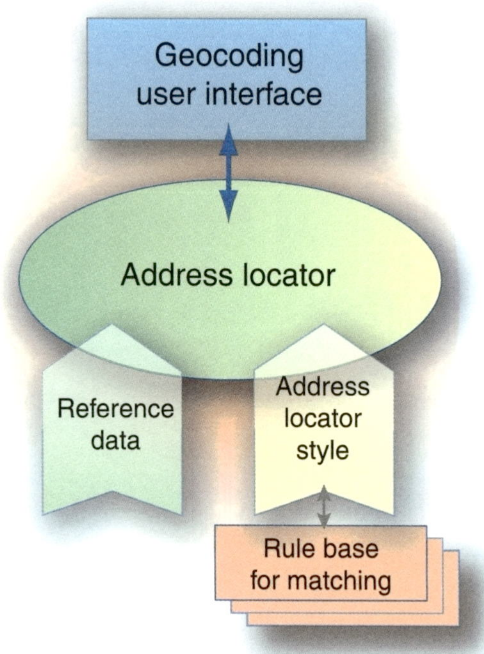

Geocoding user interface

Perhaps the most familiar of the geocoding components is the geocoding user interface. ArcGIS consists of two major applications, ArcMap and ArcCatalog. Many of the modifications and functions of geocoding can be performed through dialog boxes or tools available in ArcMap and ArcCatalog. The primary dialog boxes used in geocoding are those for creating, adjusting, and executing the geocoding process. Creation of and adjustments to an address locator are made in ArcCatalog through the Address Locator Properties dialog box. Execution can be done in either ArcMap or ArcCatalog through a variety of search-related dialog boxes. Some processes can also be performed using the toolboxes available in both ArcMap and ArcCatalog. These dialog boxes, in turn, communicate with the locator framework.

Address locator

An address locator is the combination of location-specific reference data and certain style-specific guidelines based on the address locator style selected. An address locator is created with the ArcGIS interface. This address locator is the entity that specifies the method to interpret a particular type of address input, relate it with the predefined reference data, and deliver a certain type of output back to the user interface.

The address locator style is the skeleton of the address locator. One primary file contains the adjustments that are set on the Address Locator Properties dialog box as well as pathways to use for specific rules in the rule base and reference data when performing a geocoding function. These pathways are determined by the type of address data being searched and the file formats of the reference data.

Reference data can have many different characteristics. Each address locator requires at least a primary reference dataset. The nature of this reference dataset can vary based on the entities that you wish to geocode. They are described in further detail in the chapter 'Preparing for geocoding' in this book. Beyond the Primary reference data, ancillary data can also be applied. Tables containing the place names or aliases of particular features, such as schools, government buildings, or hospitals, can be added as reference data. This allows you to search for locations based on the name of the location instead of using the street address. Alternative name tables can also be used. These tables contain lists of features, such as streets, and alternative names used for those locations. For example, a road could be known by an older name as well as a new name. A road could also be known as a highway number as well as a street name. When using an alternative name table, both names for the same feature can be searched.

Rule base

The rule base is a collection of files used to translate the address data into the desired output. For each address locator style, there are specific files in the rule base used in this translation to find attributes in the reference data that match the address attributes. These files are described extensively in the book *The Geocoding Rule Base Developer Guide*.

The process of geocoding

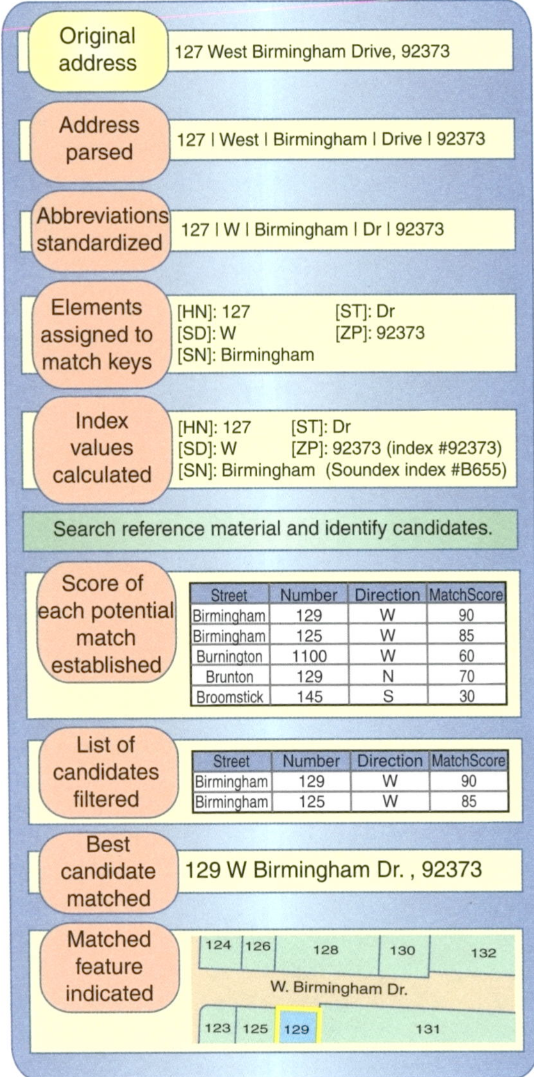

Once you have created an address locator, you can begin using it to geocode addresses. However, understanding how an address locator prepares the address data, searches the reference data, and matches addresses, as well as how modifying an address locator's settings affects this process, can help you improve both the performance and accuracy of your geocoding search.

A brief description of the geocoding process is provided below. A more detailed account is included in the chapter 'Locating addresses' in this book.

Address parsed

When an address locator standardizes an address, it dissects the address into its address elements based on the style of the address locator. Each style of address locator standardizes an address into a distinct set of address elements required for the geocoding process.

Abbreviations standardized

Many elements of an address, such as direction or street type, are often written using an abbreviation. These abbreviations are presented in a variety of formats. For example, 'Avenue' could be abbreviated as 'AV' or 'AVE'. It could also be spelled out completely as 'Avenue'. Attempting to geocode using such a wide range of abbreviations would greatly increase the search time. To avoid this, the address elements that are often abbreviated are assigned a standardized value. For example, the address elements 'West' and 'Drive' are standardized to 'W' and 'Dr', respectively.

Address elements assigned to match keys

Each element of the address being searched is assigned to a particular category, referred to as a match key. These match keys are used to compare the address data elements with the corresponding elements in the reference data.

Index values calculated

When searching for an address, some elements of the address are assigned index values that match those created in the geocoding index. These index values are used to match the address attributes to fields in the geocoding index. For example, the street name is assigned a value based on the same alphanumeric index value, known as Soundex, created for the reference data index. These index values are based on specific letters present in the street name. ZIP Codes and other zone fields are also assigned index values. These index values greatly speed up the search process.

Reference data searched

Once the address has been standardized, the address locator searches the reference data to find features with address elements that are similar to the elements of the standardized address.

Score of each potential match established

When the address locator has generated a set of potential candidates, it scores each candidate in order to determine how closely each one matches the address that you are geocoding. These scores are based on the settings established when creating the address locator. These settings include the spelling sensitivity and minimum match score and are discussed in further detail in the chapter 'Building an address locator' in this book.

List of candidates filtered

Once each candidate is scored, the address locator generates a set of candidates that are potential matches for the address. This determination is based on the minimum candidate score set for the address locator. Further details on the minimum candidate score are provided in the chapter 'Building an address locator' in this book.

Best candidate matched

The address locator finds the candidates with the highest score and produces the output format specified via the dialog boxes used when beginning the geocoding process.

Matched feature indicated

As the final step in the geocoding process, the feature that corresponds to the best candidate is indicated in the feature class, thus permitting other geoprocessing or cartographic tasks to be performed.

Customizing your address locator

There are four basic techniques for customizing your address locator. These techniques are used in conjunction with improving reference and address data to improve the amount or accuracy of your geocoding address matches. The method and application of each of these techniques are outlined below.

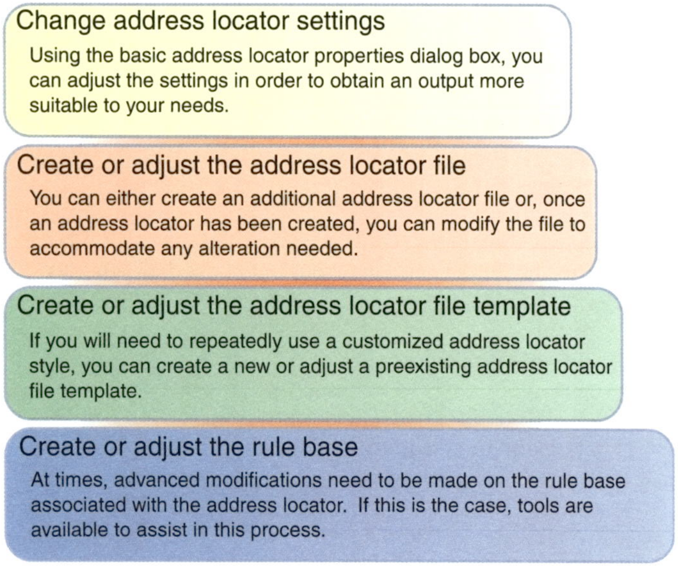

Change address locator settings

Using the basic address locator properties dialog box, you can adjust the settings in order to obtain an output more suitable to your needs.

Create or adjust the address locator file

You can either create an additional address locator file or, once an address locator has been created, you can modify the file to accommodate any alteration needed.

Create or adjust the address locator file template

If you will need to repeatedly use a customized address locator style, you can create a new or adjust a preexisting address locator file template.

Create or adjust the rule base

At times, advanced modifications need to be made on the rule base associated with the address locator. If this is the case, tools are available to assist in this process.

There is a range of methods allowing you to customize your address locator.

Changing address locator settings

There is a wide range of settings that can be adjusted on the Address Locator Properties dialog box. These range from the ability to add alternative street names and location alias names to altering the match score and spelling sensitivity settings. Making this type of adjustment is commonplace in the standard geocoding work flow. These modifications account for minor discrepancies in the reference and address data. Modifications and settings on the Address Locator Properties dialog box are described in detail in the chapter 'Building an address locator' in this book.

Creating or adjusting the address locator file

While the Address Locator Properties dialog box provides the method of making many adjustments to the individual address locator, you might find that some properties of the address locator need modification beyond that which can be done on the properties dialog box. This might include adding new geocoding styles, accommodating third party geocoding engines, using alternative data models, or altering the indexing process. Details pertaining to the related files and their settings are discussed in the chapter 'Modifying your address locator' in this book.

Creating or adjusting the address locator file template

If you find that the modifications that you made to the individual locator file would be applicable to many address locators, you may wish to create a template for the modified locator file. Many of these techniques are the same as you would apply to an individual address locator file and are also discussed in the chapter 'Modifying your address locator' in this book.

Creating or adjusting the rule base

Altering the rule base is the most advanced style of geocoding modifications. This technique is most applicable when attempting to geocode international addresses where the general address format is presented differently. Information on adjusting the rule base can be found in the book *Geocoding Rule Base Developer Guide*.

Quick reference guide to common geocoding tasks

If you would like to	Refer to
Build or maintain reference data	
Choose appropriate reference data.	• Geocoding in ArcGIS Chapter 4: 'Preparing for geocoding'
Prepare reference data.	• Geocoding in ArcGIS Chapter 4: 'Preparing for geocoding' • Editing in ArcMap
Standardize your reference data.	• Geocoding in ArcGIS Chapter 4: 'Preparing for geocoding'
Determine Address locator style	
Choose a locator style.	• Geocoding in ArcGIS Chapter 4: 'Preparing for geocoding'
Build address locator	
Build an address locator.	• Geocoding in ArcGIS Chapter 5: 'Building an address locator'
Create a dynamic feature class.	• Geocoding in ArcGIS Chapter 5: 'Building an address locator'
Locate addresses	
Prepare addresses for geocoding.	• Geocoding in ArcGIS Chapter 4: 'Preparing for geocoding ' • Editing in ArcMap
Add and remove address locators from an ArcMap document.	• Geocoding in ArcGIS Chapter 5: 'Building an address locator'
Geocode a single address.	• Geocoding in ArcGIS Chapter 6: 'Locating addresses'
Geocode a table of addresses.	• Geocoding in ArcGIS Chapter 6: 'Locating addresses'
Review and modify geocoding results.	• Geocoding in ArcGIS Chapter 6: 'Locating addresses'
Search for an intersection.	• Geocoding in ArcGIS Chapter 7: 'Additional geocoding techniques'
Use alternative street names in a search.	• Geocoding in ArcGIS Chapter 7: 'Additional geocoding techniques'
Search based on place names.	• Geocoding in ArcGIS Chapter 7: 'Additional geocoding techniques'
Publish, maintain, or customize your address locator	
Work with geocoding Web services.	• Geocoding in ArcGIS Chapter 9: 'Distributing your address locator'
Publish your geocoding service.	• Geocoding in ArcGIS Chapter 9: 'Distributing your address locator'
Edit reference data.	• Geocoding in ArcGIS Chapter 8: 'Keeping an address locator current' • Editing in ArcMap
Maintain reference data.	• Geocoding in ArcGIS Chapter 8: 'Keeping an address locator current' • Editing in ArcMap
Update your geocoding indexes.	• Geocoding in ArcGIS Chapter 8: 'Keeping an address locator current'
Use an address data model.	• support.esri.com
Modify the address locator settings.	• Geocoding in ArcGIS Chapter 5: 'Building an address locator'
Customize your address locator file.	• Geocoding in ArcGIS Chapter 10: 'Modifying your address locator'
Modify the geocoding rule base.	• Geocoding Rule Base Developer Guide

Preparing for geocoding

4

IN THIS CHAPTER

- **Understanding address locator styles**
- **Commonly used address locator styles**
- **Preparing reference and address data**
- **Standardizing your reference data**

The preparation that you do before geocoding will greatly influence the final results. This chapter discusses several of the considerations to make during this phase of the geocoding work flow.

One of the most important things to consider in building an address locator is the choice of style upon which you wish to build your address locator. An address locator style is the template on which an address locator is built. Each template is designed to accommodate a specific format of address and reference data. The address locator style template file is distinguished with a .lot file extension.

This choice of address locator styles is based on the entities you will search as well as the type of data you have. The goal of this chapter is to introduce you to the general concepts of address locator styles and help you choose the most appropriate address locator style for your application. This chapter also discusses steps you take in preparing reference and address data for geocoding.

Understanding address locator styles

Styles and the geocoding framework

In the chapter 'Concepts of geocoding' in this book, the different components of the geocoding framework were introduced. The address locator style plays an important role in the framework.

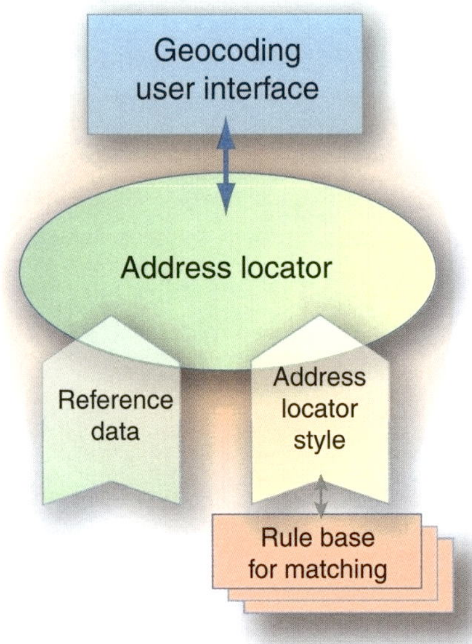

The address locator style, as a skeleton for the address locator, directs the paths and functions for the entire geocoding process.

The address locator style is the skeleton of the address locator. These files are created from one of several template files. Once created, an address locator file contains the adjustments that are set on the Address Locator Properties dialog box, the address elements to search for in the address data, and the pathways to use when performing a geocoding search. These pathways dictate the location of the reference address dataset and the address format being searched. It then directs the geocoding engine to the correct files to use for the specified address format in the rule base.

Searching for specific address elements

As described in the chapter 'Concepts of geocoding' in this book, addresses consist of specific components referred to as address elements. Address elements are arranged in a wide range of address formats. The range of elements that can be searched also varies, from individual addresses to cities to postal codes to virtually any point, line, or polygon.

Each address locator style requires that the address information being searched is presented in a specific format. Further, some styles require that the address elements be separated into individual address fields in a table, while other styles will parse the address into its individual elements automatically.

Style basics

ArcGIS Desktop comes with several predefined address locator styles that you can use immediately to create address locators. These address locator styles cover some of the most common styles of addresses that you might want to geocode. Each address locator style has specific requirements for the reference data that it can use to match addresses as well as the format of address data being searched.

You can use ArcGIS StreetMap™ data, feature classes, shapefiles, and tables as reference data for address locators. When you use feature classes and tables as reference data, they may contain some common address elements that can be used for geocoding. These elements include:

- Prefix direction (a direction that precedes the street name), as in "W. Redlands Blvd."
- Prefix type (a street type that precedes the street name), as in "Avenue B"
- Street name
- Street type (a street type that follows the street name), as in "New York St."
- Suffix direction (a direction that follows the street name), as in "Bridge St. W."
- Zone (additional information used to resolve ambiguity between addresses by identifying a region in which the address is located), as in a ZIP Code or city name.

Each address locator style has its own requirements for reference data that it can use. Each style also has different requirements for the information that tables of addresses must contain in order to be geocoded. For each style of address locator provided with ArcGIS, the requirements for reference and address data are described in this chapter.

Commonly used address locator styles

This section has been designed to help you choose the best address locator style for creating your address locator. When choosing an address locator style upon which to build your address locator, several things should be considered. This includes the type of geometry in your reference data and the format of address data you wish to search. These considerations are based on matching the address data and the reference address dataset and base files together to match an address locator style. The following table presents some of the basic characteristics of each of the address locator styles provided with ArcGIS. More in-depth descriptions are provided on the following pages.

Styles	Typical reference dataset geometry	Typical reference dataset representation	Address search parameters	Examples
US Streets	Lines	Address range for both sides of street segment	All address elements in single field	Finding a house on a specific side of the street
US Alphanumeric	Lines	Address with grid zone information	All address elements in single field	Used in some regions of Illinois and Wisconsin
US Hyphenated	Lines	Cross street information in address	All address elements in single field	Used in locations such as Queens, New York
US One Range	Lines	One range for each street segment	All address elements in single field	Finding a house on a street where side is not needed
US One Address	Points or Polygons	Each feature represents one address	All address elements in single field	Finding parcels, buildings, or address points
Single Field	Points or Polygons	Varies	Single, user-defined variable	Wide range of applications
US Cities with State	Points or Polygons	City within a state	City name and state name	Finding a specific city
World Cities	Points or Polygons	City within a country	City name and country name	Finding a specific city
ZIP 5 Digit	Points or Polygons	ZIP Code region or centroid	Five-digit postal code	Finding a specific ZIP Code location
ZIP + 4	Points or Polygons	ZIP + 4 region or centroid	Five-digit ZIP Codes and four-digit extention in separate field	Finding a specific ZIP Code location
ZIP + 4 Range	Points or Polygons	Region covering several ZIP + 4 ranges	Five-digit ZIP Codes and four-digit extention in separate field	Finding a more general location for each ZIP + 4

US Streets

The US Streets address locator style lets you create address locators for common addresses. One advantage of this address locator style is that it permits you to provide a range of house number values for both sides of a street segment. With this, the address locator can deliver not only a location along the street segment, but it can also determine the side of the road segment where the address is located. For example, in the illustration below, in order to find the address 750 Juniper St. NE, you would determine that it is not only located on the street segment near the middle of the illustration, but you would also be able to determine that it is on the east side of the street.

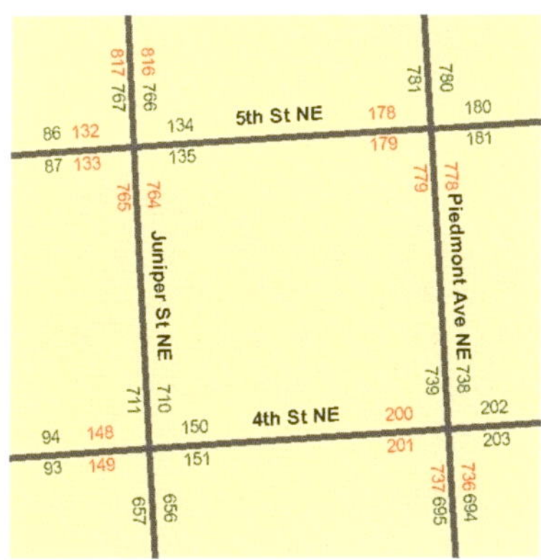

When using the US Streets address locator style, each road segment has a to and from address range for both the right and left side of the road.

This address locator style can use feature classes with any type of geometry but typically uses feature classes with line geometry. Each feature in the reference data represents a street segment with two ranges of addresses that fall along that street segment, one for each side of the street.

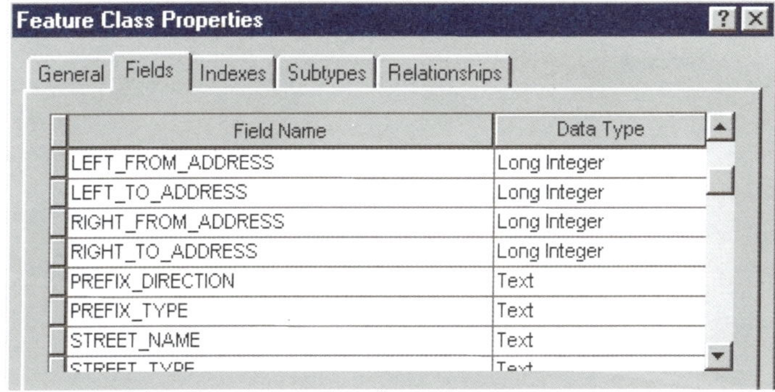

Feature Class Properties

General | Fields | Indexes | Subtypes | Relationships

Field Name	Data Type
LEFT_FROM_ADDRESS	Long Integer
LEFT_TO_ADDRESS	Long Integer
RIGHT_FROM_ADDRESS	Long Integer
RIGHT_TO_ADDRESS	Long Integer
PREFIX_DIRECTION	Text
PREFIX_TYPE	Text
STREET_NAME	Text

The reference data attribute table used in the US Streets address locator style must contain fields pertaining to the to and from address values for both the left and right sides of the road.

To use a shapefile or feature class as reference data for a US Streets style of address locator, it must have fields that contain from address and to address information for each side of the street, and street name information as well as an ObjectID field and SHAPE field. Optionally, you can specify fields that contain the street's prefix direction, prefix type, street type, suffix direction, or zone.

Each of the U.S. address locator styles, while having different requirements for reference data, has the same requirements for input address data. Tables of addresses that can be geocoded using these address locators must have an address field containing the street number and street name in addition to the street's prefix direction, prefix type, street type, or suffix direction, if any. Intersection descriptions (for example, "Eureka Blvd. & Vine St.") can also be included in this field.

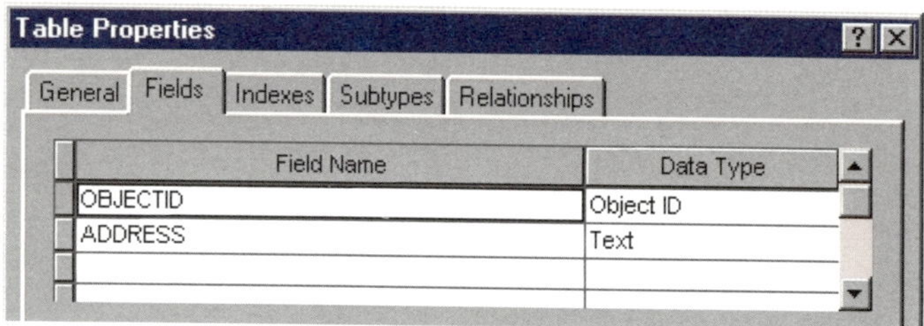

For all US address locator styles, the address table must contain a field with the address information. The address elements do not need to be broken up into individual fields.

US Alphanumeric Ranges

The US Alphanumeric Ranges address locator style lets you create address locators for U.S. addresses that contain alphanumeric house number ranges. Such alphanumeric house ranges are used in some regions of Wisconsin and Illinois. The additional values usually represent a grid zone value. For example, the address N84W 16301 W Donald Ave suggests that the address is not only at 16301 W Donald Avenue, but that it is also in grid zone N84W.

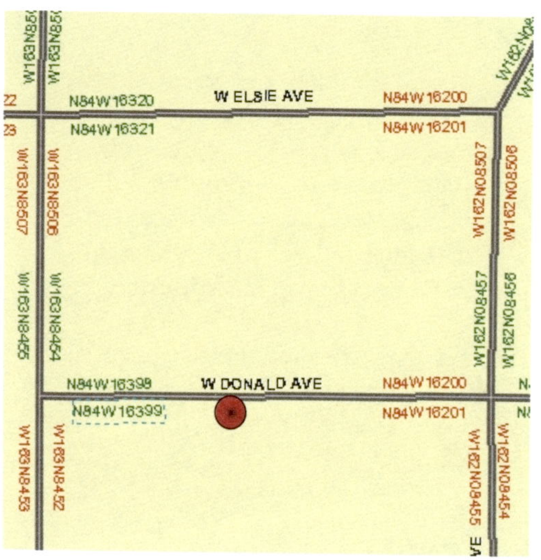

Alphanumeric house ranges include those address formats that have the grid zone included with the house number.

This address locator style can use feature classes with any type of geometry but typically uses those with line geometry representing the street network. Each feature in the reference data represents a street segment with two ranges of addresses that fall along each street segment, one for each side of the street.

To use a feature class as reference data for a US Alphanumeric Ranges style of address locator, it must have fields that contain a grid zone, from address and to address information for each side of the street, street name information, an ObjectID field, and a SHAPE field. In addition, you can specify fields that contain the street's prefix direction, prefix type, street type, suffix direction, or zone.

Feature Class Properties

General | Fields | Indexes | Subtypes | Relationships

Field Name	Data Type
GRIDZONE	Text
LFROM	Long Integer
LTO	Long Integer
RFROM	Long Integer
RTO	Long Integer
PREFIX	Text
NAME	Text
TYPE	Text

When using the US Alphanumeric Ranges address locator style, the reference data must contain fields for the grid zone as well as the to and from address values for both sides of the street. Additional address elements are also required.

The US Alphanumeric Ranges address locator requires the grid zone and the house number ranges to be stored in different fields. For example, the address ranges N95W19800–N95W19898, N95W19801–N95W19899 need to be broken down into separate fields as shown in the following illustration.

GRIDZONE	LFROM	LTO	RFROM	RTO	PREFIX	NAME	TYPE
N95W	17539	17715	17538	17714	W	SHADY	LN
W203N	9599	9401	9598	9400	N	BITTERSWEET	LN
N95W	19800	19898	19801	19899	W	AZALEA	RD
N95W	19098	18900	19099	18901	W	CINDY	CIR
W166N	9501	9557	9500	9556	N	RICHMOND	DR
W167N	9401	9533	9500	9564	N	CUMBERLAND	RD
N95W	17391	17537	17390	17536	W	SHADY	LN
W188N	9598	9508	9599	9509	N	MAPLE	RD
W161N	9401	9599	9400	9598	N	BLACKFOOT	DR

When viewing the attribute table of the feature class used for a US Alphanumeric Ranges address locator style, the grid zone information has been placed in its own field, while the house number information is recorded in four fields including the right to and from as well as the left to and from address ranges.

The US Alphanumeric Ranges address locator style has the same address table requirements as the US Streets address locator style.

US Hyphenated Ranges

The US Hyphenated Ranges address locator styles lets you create address locators for U.S. addresses that contain hyphenated house number ranges. The hyphenated ranges depict a number that is usually the number of the cross street, followed by a hyphen, then the actual number of the house along the street (for example, 76-20 34th Ave). One location that uses this type of address style is Queens, New York. The first number indicates either the north or west cross street. The second number indicates where on the block the building is located.

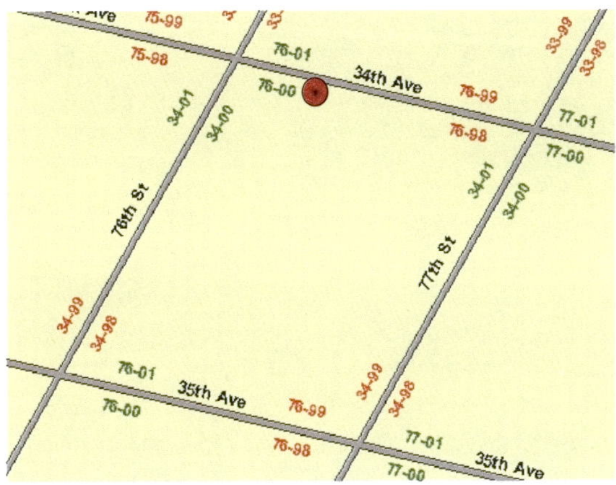

The US Hyphenated Ranges address locator style is used for addresses that contain a hyphenated house number; the additional number usually represents the nearest cross street.

This address locator style can use feature classes in a geodatabase with any type of geometry but typically uses feature classes with line or polyline geometry. Each feature in the reference data represents a street segment with two ranges of addresses that fall along that street segment, one for each side of the street. The US Hyphenated Ranges address locator style only supports feature classes in a geodatabase.

Feature Class Properties

General | Fields | Indexes | Subtypes | Relationships

Field Name	Data Type
LFROM	Text
LTO	Text
RFROM	Text
RTO	Text
PREFIX	Text
PRE_TYPE	Text
NAME	Text
TYPE	Text

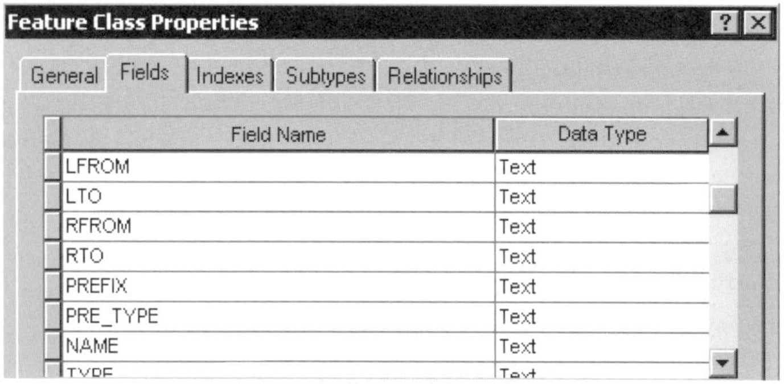

The attribute table for the reference data being used in a US Hyphenated Ranges address locator must contain individual fields for the address ranges for the left and right sides of the road segment. These can include hyphenated entries or simply the house numbers. Other fields are also required.

To use a feature class as reference data for a US Hyphenated Ranges style of address locator, the related attribute table must have fields that contain from address and to address information for each side of the street, street name information, an ObjectID field, and a SHAPE field. You can also specify fields that contain the street's prefix direction, prefix type, street type, suffix direction, or zone.

In the from address and to address fields of the feature class, the house number can be a hyphenated number or a simple house number. For example, as shown in the following table, the address ranges 75-01–75-99, 75-00–75-98 must contain a hyphen that separates the cross street and the actual house number. A simple house number range, such as 101–199, can be used if the range is not a hyphenated range.

LFROM	LTO	RFROM	RTO	PREFIX	PRE_TYP	NAME	TYPE	SUF
75-01	75-99	75-00	75-98			35th	Ave	
76-01	76-99	76-00	76-98			35th	Ave	
77-01	77-99	77-00	77-98			35th	Ave	
33-01	33-99	33-00	33-98			76th	St	
101	199	102	198			Main	St	
75-01	75-99	75-00	75-98			34th	Ave	
34-01	34-99	34-00	34-98			76th	St	
76-01	76-99	76-00	76-98			34th	Ave	

The attribute table related to the reference data may contain address range values that consist of a hyphenated address, consisting of the cross street and house number, or simply the house number if no cross street information is included in the address.

When a geocoding index is built for the feature class using the US Hyphenated Ranges address locator style, the cross street and house number information will be separated in different fields in the index table as shown in the image above. The house number of an input address will be matched against the house number fields in the index table. For more information on geocoding indexes, see the chapter 'Keeping an address locator current' in this book.

The US Hyphenated Ranges address locator style has the same address table requirements as the US Streets address locator style.

US One Range

The US One Range style lets you create an address locator for street segments with address ranges. This address locator style is quite similar to the US Addresses address locator; however, this style only requires one range for each road segment. This data is less extensive and can be used when the side of the street in the output data is not essential. This address locator style typically uses feature classes with line or polyline geometry. Each feature in the reference data represents a street segment with a range of addresses that fall along that street segment.

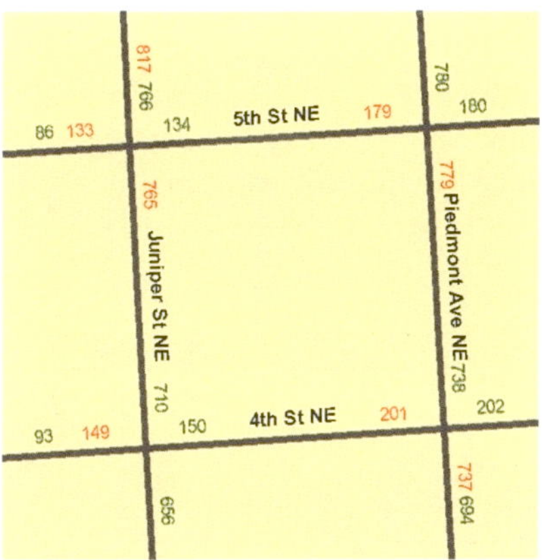

When using the US One Range address locator style, each road segment must have a to and from value.

To use a feature class as reference data for a US One Range address locator, it must have fields that contain from address, to address, and street name information in addition to an ObjectID field and a SHAPE field. In addition, you can optionally specify fields that contain the street's prefix direction, prefix type, street type, suffix direction, or zone. The use of zones will be discussed later in this chapter.

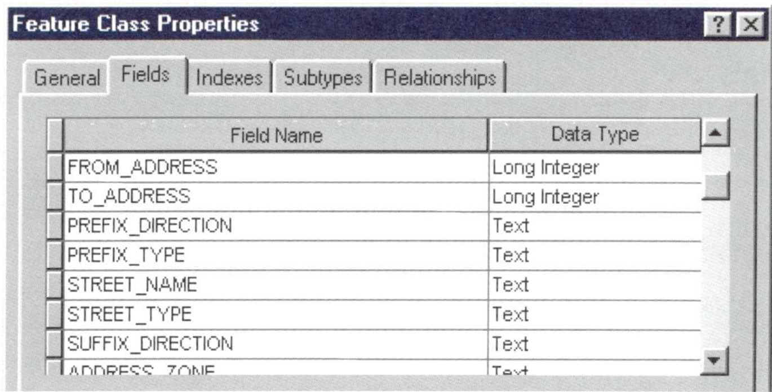

Reference data attribute fields for the US One Range address locator style must include the from and to address as well as the street name for each segment of road.

The US One Range address locator style has the same address table requirements as the US Streets address locator style.

US One Address

The US One Address locator style lets you create address locators for United States addresses. US One Address locators use feature classes with polygon or point geometry as reference data. Each feature in the reference data corresponds to a single address. For example, you could use a feature class containing parcel polygons, building footprints, or parcel centroids (the center points of parcel polygons) as reference data for a US One Address locator. Each address that you wish to search must be present on the reference data. Exact locations cannot be extrapolated from any type of range of addresses on a street.

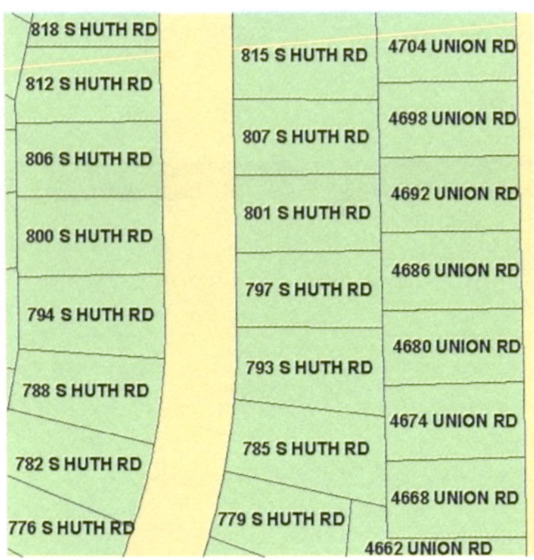

When using the US One Address address locator style, each feature in the reference data must correspond to a single address value such as parcels or buildings.

To use a feature class as reference data for a US One Address locator, it must have individual fields that contain a street number and street name information in addition to an ObjectID field and a SHAPE field. Beyond this, you can specify additional fields that contain the street's prefix direction, prefix type, street type, suffix direction, or zone.

The US One Address locator style has the same address table requirements as the US Addresses locator style.

Single Field

The Single Field address locator style lets you create address locators for address data that contains the location information in a single field. This address locator style has a broad range of applications. You could use a Single Field address locator style to geocode location descriptions such as place names, city names, or state names. Beyond that, you could use the Single Field address locator style to create address locators to find hydrologic

Single-field addresses include features with one address element in an individual field.

units, census tracks, and virtually any unique feature represented in a feature class. More information on these types of address locators is offered in the chapter 'Additional geocoding techniques' in this book.

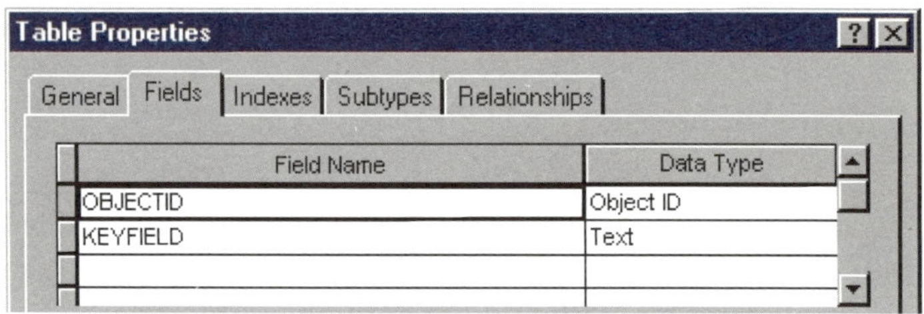

Tables of addresses being geocoded using the Single Field address locator style must contain a single address field that corresponds to the unique address field in the reference data. This field can contain features ranging from cities to hydrologic units.

Although Single Field address locators can use feature classes with any type of geometry, they typically use feature classes with point or polygon geometry as reference data. In addition to an ObjectID field and SHAPE field, feature classes that you can use as reference data for a Single Field address locator must have a key field that contains the unique "address" for that feature. It is this field in the attribute table that is referenced when geocoding against the reference data.

Tables of addresses that can be geocoded using a Single Field style of address locators must also contain a single field with the same unique "address" information that can be used to identify the locations.

US Cities with State

The US Cities with State address locator style lets you create address locators for city names that contain fields that have city and state name information. This address locator style can use feature classes with point or polygon geometry. Each feature in the reference data represents a city polygon or its centroid.

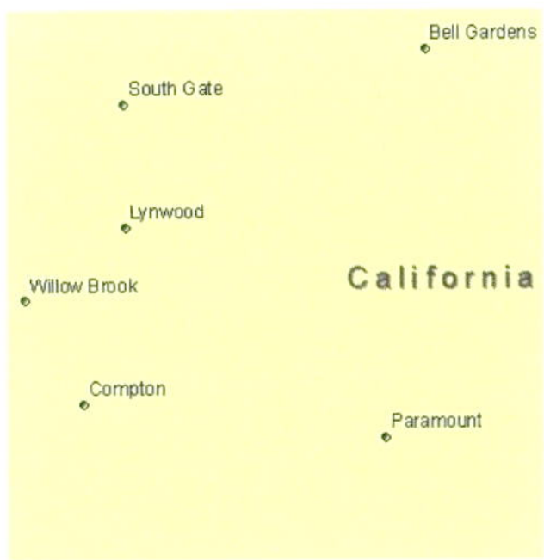

You can create an address locator to locate cities within a state using the US Cities with State address locator.

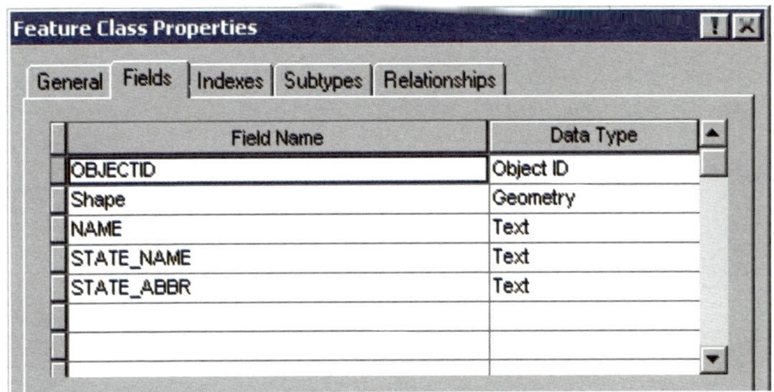

Field Name	Data Type
OBJECTID	Object ID
Shape	Geometry
NAME	Text
STATE_NAME	Text
STATE_ABBR	Text

When using the US Cities with State address locator style, the attribute field must contain the name of the city as well as the state information, either the name or the abbreviation.

Reference data for a US Cities with State style of address locator must have fields that specify the city name and the state name or its abbreviation for the feature, in addition to an ObjectID field and a SHAPE field.

Tables of addresses that can be geocoded using these address locators must also contain fields that have city name and state-specific information.

World Cities

The World Cities with Country address locator style lets you create address locators for city names that contain the location name information in two fields such as city and country. This address locator style can use feature classes with point or polygon geometry. Each feature in the reference data represents a city polygon or its centroid.

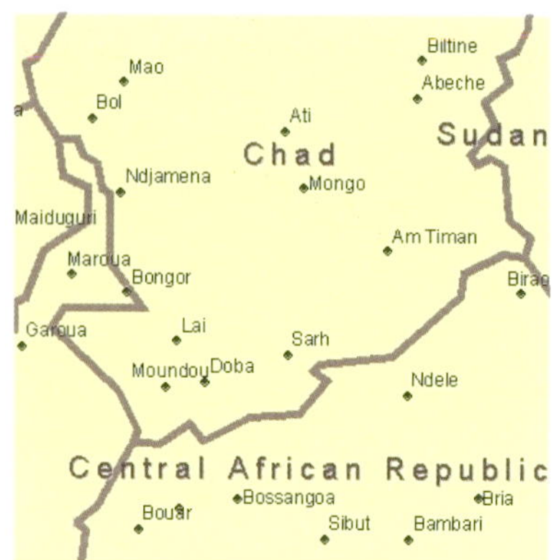

With the World Cities address locator style, you can create an address locator to locate international cities.

Reference data for a World Cities with Country style of address locator must have fields that specify the city name and the country name for the feature in addition to an ObjectID field and a SHAPE field.

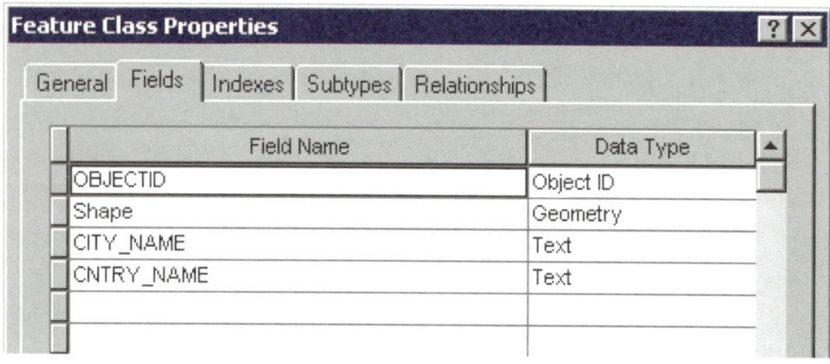

Field Name	Data Type
OBJECTID	Object ID
Shape	Geometry
CITY_NAME	Text
CNTRY_NAME	Text

When using the World Cities address locator style, the attribute field must contain the name of the city as well as the country name.

Tables of addresses that can be geocoded using these address locators must also contain a city name field and a country name field.

ZIP 5 Digit

The ZIP 5 Digit address locator style lets you create address locators for postal codes. While specifically designed for U.S. five-digit ZIP Code values, any short integer postal code could be used. This address locator style uses feature classes with point or polygon geometry. Each feature in the reference data represents a ZIP polygon or its centroid.

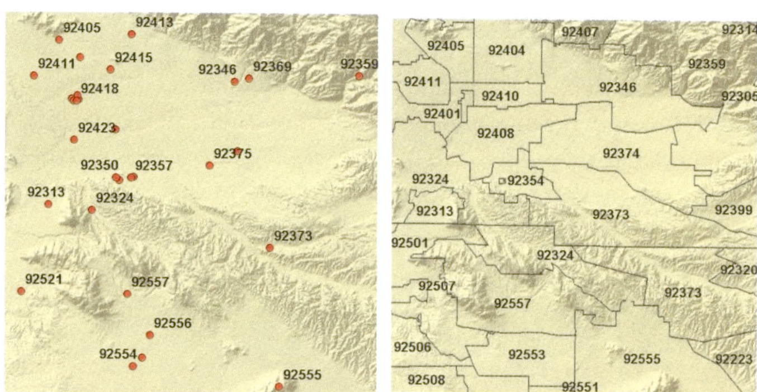

Your reference data for a ZIP 5 Digit address locator can consist of points or polygons representing postal regions or their centroids.

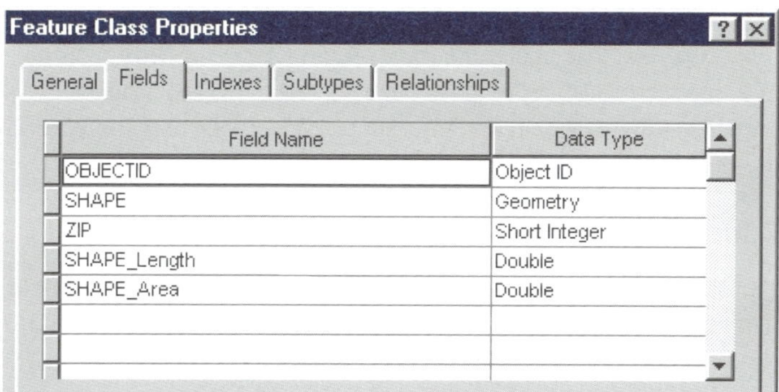

The reference data attribute table for a ZIP 5 Digit address locator must contain a field specifying the ZIP Code of the point or polygon.

Reference data for a ZIP 5 Digit style address locator must have a field that specifies the short integer postal code for the feature in addition to an ObjectID field and a SHAPE field.

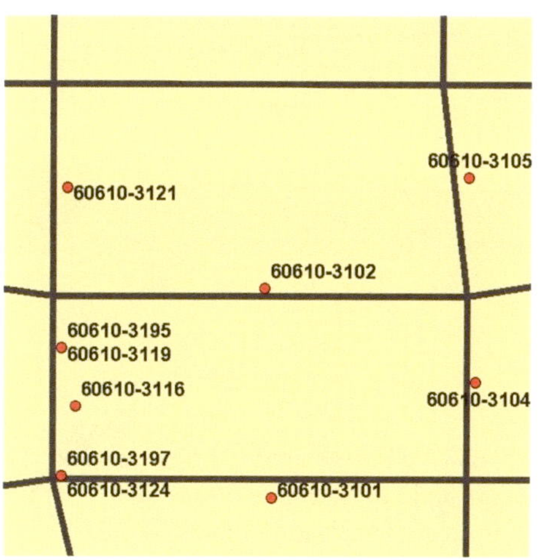

With the introduction of the four-digit suffix added to the ZIP Code, an address locator can pinpoint the location with greater accuracy.

Tables of addresses that can be geocoded using these address locators must contain a field that has short integer postal code information.

ZIP + 4

The ZIP + 4 address locator style has been specifically designed for United States ZIP + 4 codes. This address locator style can use feature classes or shapefiles with point or polygon geometry. Each feature in the reference data represents a ZIP + 4 polygon or its centroid.

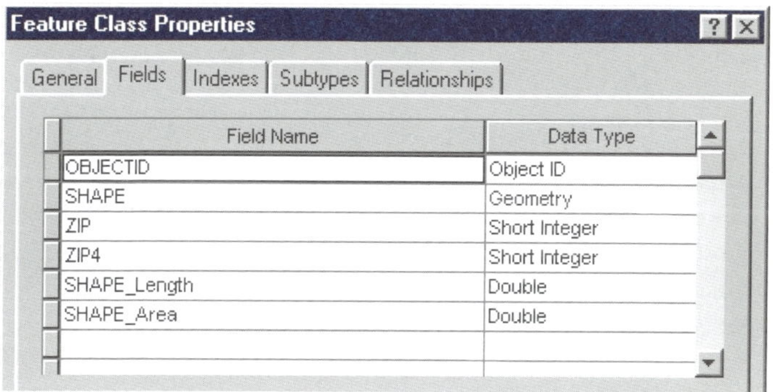

The reference data attribute table used in a ZIP + 4 address locator must contain a field specifying the standard five-digit ZIP Code and an additional field containing the additional four-digit suffix value.

Reference data for a ZIP + 4 style of address locator must have a field that specifies the five-digit ZIP for the feature and a field containing the four-digit add-on code for the feature in addition to an ObjectID field and a SHAPE field.

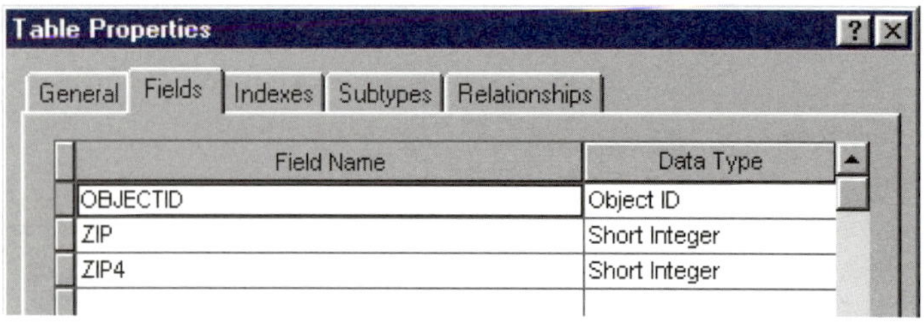

An address table must contain the ZIP field as well as a field containing the four-digit suffix in order to be geocoded using the ZIP + 4 address locator.

Tables of addresses that can be geocoded using these address locators must contain a field that has five-digit ZIP and an additional field containing the four-digit add-on code information.

ZIP + 4 Range

The ZIP + 4 Range address locator style lets you create address locators for a range of United States ZIP + 4 codes. This address locator style can use feature classes with point or polygon geometry. Each feature in the reference data represents a contiguous block with a specific range of ZIP + 4 codes.

To use a feature class or shapefile as reference data for a ZIP + 4 Range style of address locator, it must have fields that specify the five-digit ZIP for the feature and the lower and upper bounds for the four-digit add-on code in addition to an ObjectID and a SHAPE field.

The ZIP + 4 Range address locator style has the same address table requirements as the ZIP + 4 address locator style. A match is assigned to a feature that covers the range of ZIP + 4 values.

Additional style properties

As you peruse the list of address locator styles on the Create New Address Locator dialog box, you will notice that there are other additional attributes to distinguish the different address locator styles. These include the use of a shapefile termed "File" or a geodatabase feature class abbreviated to "GDB", the presence of an alternative name file referred to as "AltName", and the presence of a Zone field in the reference data.

Choosing between File and GDB

The terms File and GDB are referring to the type of reference data that will be used in the geocoding process. If a shapefile will be used, select an address locator style that ends with (file). If a feature class from a geodatabase will be used, select an address locator style that ends with (GDB).

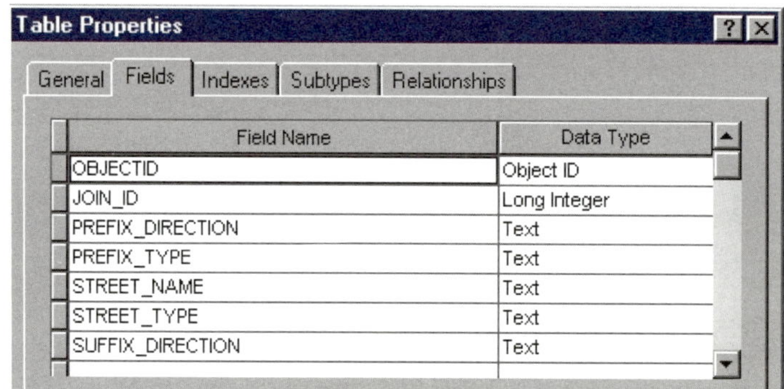

When using an alternative street name table, there are certain required fields it must contain.

Alternate street names

For the US One Address, US One Range, and US Streets address locator styles, you can use a table to define alternate street names for the features in your reference data feature class. Using alternate street names allows you to match an address to a feature using one of many names for the feature. For example, if “Bridge Street” is also known as “Slash Road”, then you can find the same address using “266 Bridge Street” or by using “266 Slash Road”.

Tables that you use to specify alternate street names must have an ID field, a JOIN_ID that specifies the feature in the reference data to which the alternate name applies, and an alternate street name field. Optionally, the table can contain fields that contain prefix direction, prefix type, street type, or suffix direction information. You can specify multiple alternate names for the same feature in your reference data by creating records in the alternate street name table with the same JOIN_ID, referencing the same feature in the reference data feature class.

Each record in an alternate street name table applies to only one feature in your reference data feature class. In order to specify an alternate street name for all features that make up a particular street in your reference data feature class, you must create a record in the alternate street name table for each feature in your reference data feature class.

Using the zone field

Many times additional fields are found on the reference data that act to further clarify the location of the attribute. This includes such entities as postal codes, states, or countries. This type of information is referred to as a zone. This additional information can be used to increase the likelihood of a correct match.

The zone attribute for the reference data is set on the Primary Table tab on the Address Locator Properties dialog box and is defined for the address data on the Find dialog box used when performing a geocoding search.

Preparing reference and address data

Beyond matching your reference and address data to the required format for the particular address locator style that you have selected, there are several other issues to keep in mind regarding your reference and address data.

Recognizing standardized values for abbreviated address elements

It is important to remember that many of the address data abbreviations are standardized into a particular format. For example, the standard abbreviation for “West” as a street direction received the complete name “West”. If your reference data has abbreviated values different from those produced in the geocoding process, such as “Wst.” or “W.”, the resulting match scores will be substantially lower. There are some tools available to standardize your reference data. These tools are described in the section ‘Standardizing your reference data’ in this chapter.

Grammatical errors

Many times the spelling of an address element in the reference data might differ from the spelling used in the address data. The misspelling of one or more elements will contribute to the lowering of the match score. It is important that the spelling used for the reference data attributes matches the parameters being searched as the address data. If an address element is spelled using a variety of methods, the use of an alternate street name table is advised. More information about alternate place name tables is provided in the chapter ‘Additional geocoding techniques’ in this book.

Incomplete reference data

In the majority of geocoding applications, the need arises from time to time for the reference data to be updated. For example, if a new housing tract is added to the city street network, these additional streets, with their associated ranges and other properties, need to be added.

If these elements are not added, when the reference data is searched for a street that falls within this region, no match will be found. Only the elements that exist in the reference data can be found via a geocoding search. This is also the case for address ranges and other characteristics related to the reference data. Additional information regarding updating reference data and maintaining your geocoding indexes can be found in the chapter ‘Keeping an address locator current’ in this book.

Standardizing your reference data

Within the geocoding toolbox, ArcGIS provides a standardize address tool to automatically standardize your reference data.

1. In ArcToolbox™, open the geocoding toolbox and double-click the Standardize Addresses tool.

   This will open the Standardize Addresses dialog box.

2. Click the Browse button next to the Input Address Data text box, and navigate to the table of addresses you wish to standardize.
3. In the Input Address Fields text box, add or remove the field names you wish to include in the output table.
4. Click the Browse button next to the Address Locator Style text box, and navigate to the address locator style you wish to use to standardize your reference data table.

   The output address field names will appear in the Output Address Fields text box.

5. Click the Browse button next to the Output Address Data text box and navigate to the folder where you wish to store the output address table.
6. In the Navigate dialog box, specify a name and file type for the output address table and click OK.
7. Click OK on the Standardize Addresses dialog box to begin the operation.

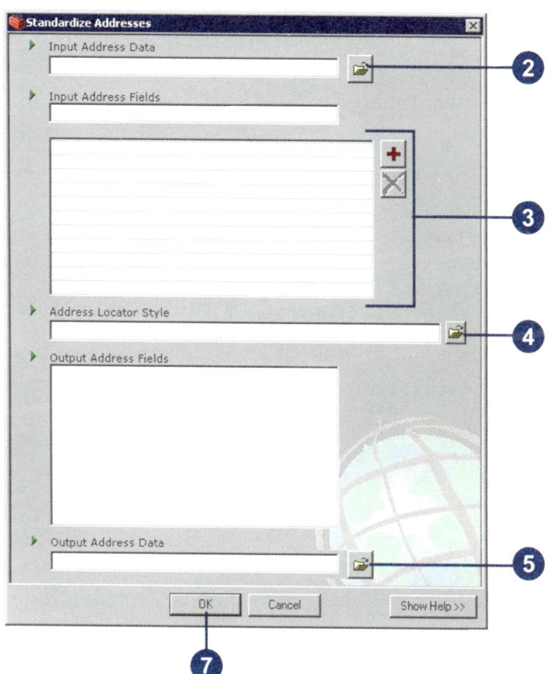

Building an address locator

5

IN THIS CHAPTER

- **Creating an address locator**
- **Modifying an address locator's settings**
- **Managing address locators in ArcMap**

The address locator is the cornerstone of the geocoding process in ArcGIS. Once you have created an address locator, you can use it in a variety of ways within ArcGIS to find addresses.

This chapter describes the process of creating an address locator in ArcGIS, modifying its properties, and other tasks related to this step in the geocoding work flow.

Creating an address locator

One of the first physical processes that you do when geocoding is the creation and tailoring of an address locator. This process always begins in ArcCatalog. The Address Locators folder at the top of the Catalog tree lets you manage existing locators and create new address locators on your computer.

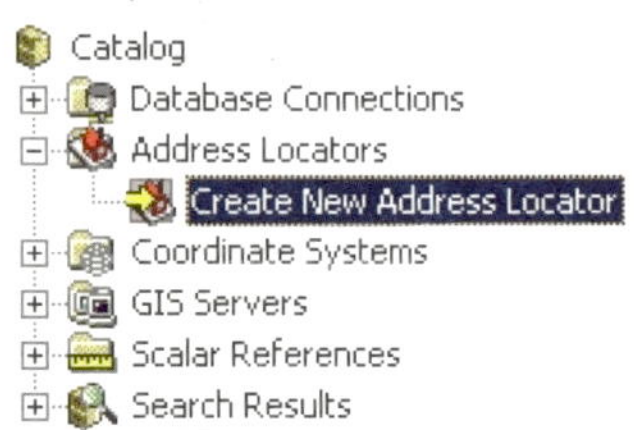

The Address Locators folders in ArcCatalog are the default locations for creating and storing locators.

Once the Create New Address locator icon in the ArcCatalog tree has been double-clicked, the Create New Address Locator dialog box will open. On this dialog box you have the option to select the address locator style that you wish to use to create the address locator. There are many different address locator styles available, and the correct selection of an address locator is based on the format of the address data, the style and format of your reference data, and the use of alternate name tables. A detailed description of each of the styles and their proper application is given in the chapter 'Preparing for geocoding' in this book.

With the selection of an address locator style, the Address Locator Properties dialog box is the next interface used in the creation of an address locator. This dialog box consists of several sections, each contributing to the performance and results of the geocoding process.

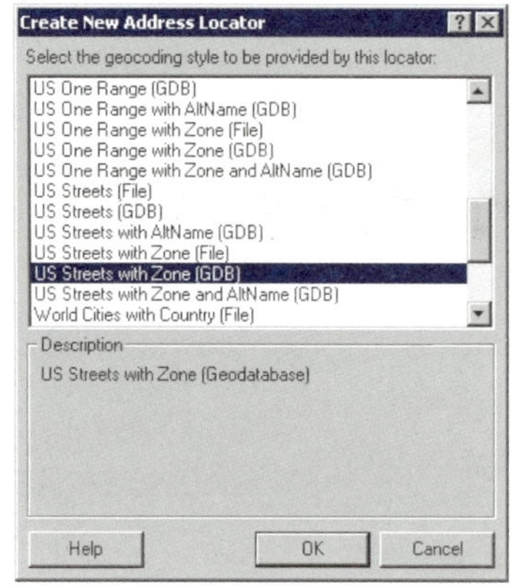

A wide range of address locator styles is available for use in the creation of an address locator. The correct selection of style is based on the address and reference data formats.

Name and description

The name that you use to identify the address locator is used in the address locator manager in ArcMap and in the Catalog tree in ArcCatalog. The name is added as a suffix to <yourLoginName>.

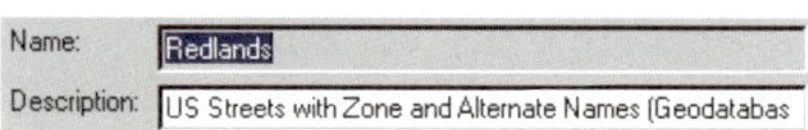

The Name and Description text boxes allow you to provide a name for, and add any additional description to, the address locator.

For example, if your login name was "ssmith", and you chose "Atlanta" as the name of your address locator, it would be stored as "ssmith.Atlanta". This helps identify the creator of the address locator. The description that you enter for your address locator allows you to provide a more detailed description of the address locator. This brief description is present whenever the Address Locator Properties dialog box is open and is listed on the Address Locator Manager dialog box in ArcMap.

Primary table

The primary table tab is the interface used to specify the primary reference data to be used by the address locator. The reference data can be either ESRI StreetMap data, a shapefile, or a feature class. The subsequent fields match the fields required for the address locator style to the attribute fields in the reference data.

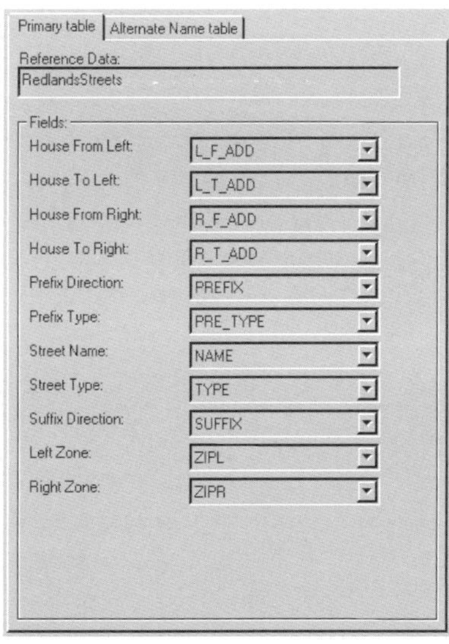

The Primary table tab is the interface where you assign a particular feature class to act as the reference data. The Alternate Name table tab is used when alternative names are available for streets or other attributes in the reference data.

Fields

Based on the service style you chose, the fields available to match will vary. By default, the address locator searches the reference attribute table and automatically matches attribute fields to the fields required by the address locator. If the default match fields are not correct, you can manually match the fields required using the dropdown menu for each field.

Alternate Name table

Often, streets or other attributes in the reference data are referred to by multiple names or the name changes over time. If this is the case, you may need to geocode addresses using alternate names.

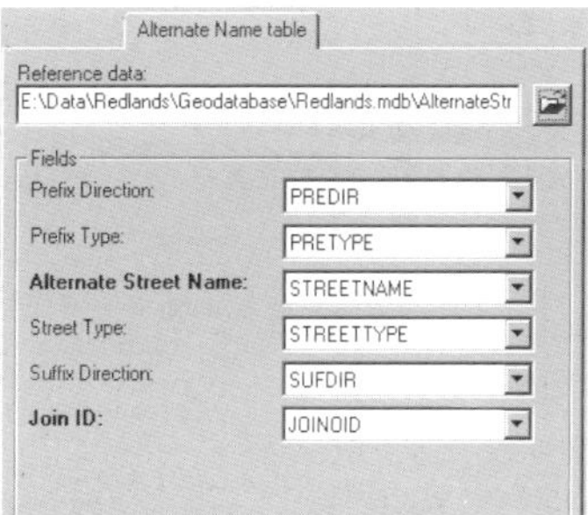

The Alternate Name table tab allows you to specify an additional table containing different names for items on your primary table.

The Alternate Name table tab allows you to specify an additional table containing different names for the items on your primary table. A more detailed description of this dialog box and its functionality is provided in the chapter 'Additional geocoding techniques' in this book.

Input address fields

When you geocode a table of addresses, an address locator will try to determine which fields in the table contain particular pieces of address information. An address locator accomplishes this by searching for specific default input address field names. When the address locator finds a field with one of the default address field names, then the address locator will automatically use the contents of that field for a particular address component.

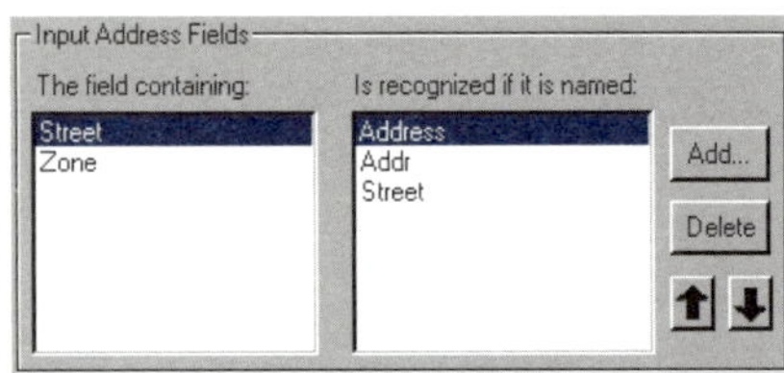

The Input Address Fields section is used by the address locator to recognize alternative field names that might be present in the address data.

You can use the input address field settings of an address locator to specify names of fields in your address tables that are likely to contain address information. For example, a field that contains the zone component of an address might be called "City", "ZIP", "ZIPCode", or "Zone". If your address locator requires zone information for addresses that it can geocode, then you can use the input address field settings to manually enter the names of the fields containing the address information. The address locator will then search for fields with these names when it tries to find the zone information in a table of addresses.

Matching options

The Matching Options section is used to set user-defined settings for the address locator. The Place Name Alias Table button opens an additional dialog box that allows you to specify a table containing alternative location names and the corresponding addresses.

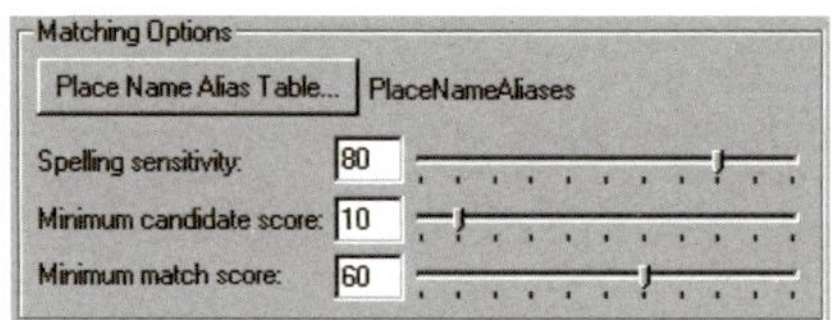

Place Name Alias Table

Depending on the geocoding application, you may wish to search for locations based on the location names, such as the name of a hospital or school. The Place Name Alias Table button opens an additional dialog box that can be used to refer to a table that contains a location name and the related address information.

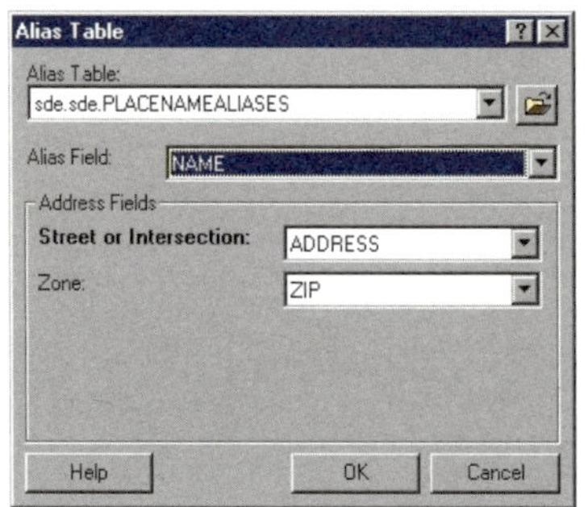

You can use an alias table to search for locations, such as a specific hospital or school, by name rather than by address.

This allows the name of the location to be provided as the search parameter, directing the address locator to then search for the location address in the reference data. A more detailed description of this dialog box and its functionality is provided in the chapter 'Additional geocoding techniques' in this book.

Spelling sensitivity

The spelling sensitivity setting controls how much variation the address locator will allow when it searches for likely candidates in the reference data. A low value for spelling sensitivity will allow "Mane", "Maine", and "Man" to be treated as match candidates for "Main". A higher value will restrict candidates to exact matches. The spelling sensitivity does not affect the match score of each candidate; it only controls how many candidates the address locator considers. The address locator then computes the match score of each candidate and ranks the candidates by score.

The spelling sensitivity setting for an address locator is a value between 0 and 100. By default, the spelling sensitivity is 80, which allows for only minor variations in spelling. If you are sure that your addresses are spelled correctly, you could set a higher spelling sensitivity. If you think that your addresses may contain spelling errors, then you should use a lower setting. The geocoding process takes longer when you use a lower setting because the address locator has to compute scores for more candidates.

Minimum candidate score

When an address locator searches for likely candidates in the reference data, it uses this threshold to determine whether a potential location should be considered. Locations that yield a match score lower than this threshold will not be considered a candidate.

The minimum candidate score for an address locator is a value between 0 and 100. By default, this is set to 30. If the address locator seems unable to find any likely candidates for an address that you want to geocode, you could lower this setting so that candidates with low scores are considered.

Minimum match score

The minimum match score setting lets you control how well addresses have to match their most likely candidate in the reference data in order to be considered matched. A perfect match yields a score of 100. A match score between 80 and 100 can generally be considered a good match. An address below the minimum match score is considered to have no match.

The minimum match score for an address locator is a value between 0 and 100. By default, this setting is 60. If your application demands that addresses be located with a high level of confidence, you should set a higher minimum match score. If you want to maximize the number of addresses that can be matched and don't mind if some addresses are potentially matched incorrectly, you can use a lower setting.

Intersection connectors

Address locators that are based on the US One Range, US Streets, and ArcView StreetMap address locator styles can geocode street intersections in addition to street addresses. In ArcGIS, intersections are designated as two streets delimited by an intersection connector string. Some examples of intersection descriptions are "Hollywood Blvd. & Vine St." and "Yonge St. and Bloor St.".

The intersection connectors setting lets you specify all of the strings that the address locator will recognize as intersection connectors. By default, "&", "|", and "@" are recognized as intersection connectors. You can add additional symbols or words to indicate intersection connectors.

When geocoding using street intersections, the Intersections section specifies the symbology that will be used to separate the street names.

Output options

The Output Options section allows you to specify the precise location on the map at which a feature will be drawn.

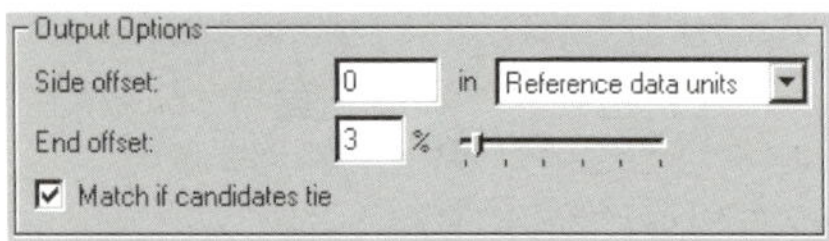

Side offset

Some styles of address locators use reference data that contains address range information for each side of the street (e.g., US Streets and StreetMap address locator styles). Address locators based on these styles can determine on which side of the street an address is located. For cartographic purposes, you can specify a side offset for geocoded features when using these styles of address locators. When you specify a side offset, the address locator locates geocoded features at the specified distance from the street centerline on the correct side of the street.

End offset

Address locators that use reference data with line geometry, such as those based on the US One Range, US Streets, or StreetMap address locator styles, can interpolate a position along reference features for a geocoded address. In order to prevent features that are located at the end of a reference feature from falling on top of

The address, 100 MAIN ST, has been offset from the street feature by the offset distance of 25 feet. This address falls at the end of the street feature and is, therefore, in line with the end of the street feature.

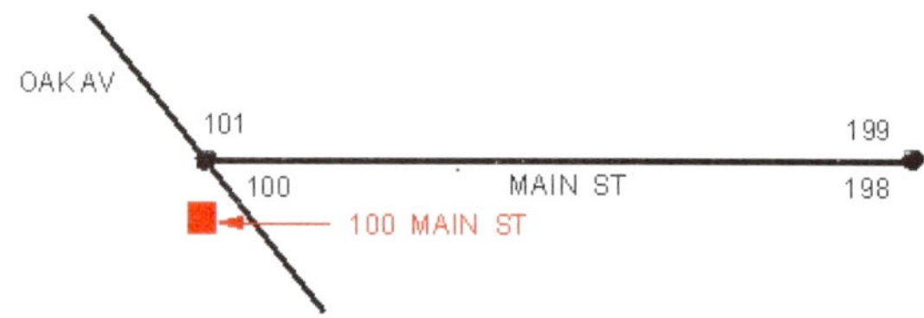

When streets intersect at odd angles, specifying an offset distance can have the undesirable effect of placing the address so it appears that the address does not belong to MAIN ST, but rather to OAK AV.

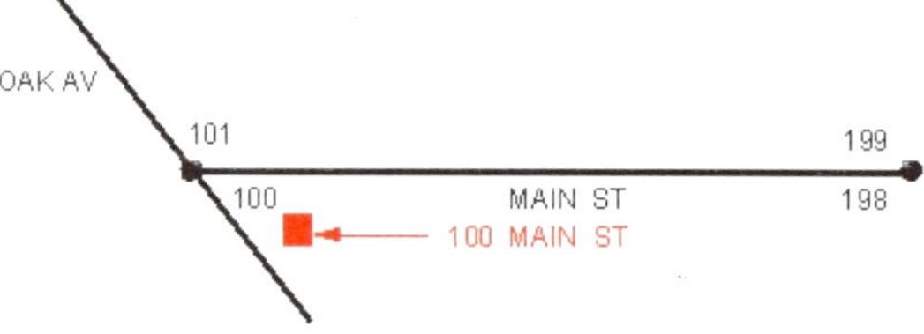

An end offset can be specified that adjusts the location of the address toward the center of the street feature. Using an end offset will often rectify the condition shown in the previous diagram. In this example, a squeeze factor of 10 percent was used to move the address toward the center of the street feature by a distance equal to 10 percent of the length of the street feature.

other features (for example, a cross street), the address locator can apply a "squeeze factor", or end offset, to the location of a geocoded address. The end offset setting of an address locator is expressed as a percentage of the length of the reference feature, between 0 and 50 percent. An end offset setting of 0 percent will not offset features from the end of the reference feature. An end offset of 50 percent will locate all features at the middle of the reference feature. By default, the end offset setting for an address locator is 3 percent.

Match if candidates tie

If an address locator finds two or more reference features that have the same highest match score, you can specify whether or not to match an address arbitrarily to one of these features. Use this setting to specify whether to arbitrarily match these addresses or to leave them unmatched. In either case, you can review addresses with tied candidates during the interactive review process whether or not they are matched. The review process is described in detail in the chapter 'Locating addresses' in this book.

Output fields

When creating a feature class with the geocoded addresses, the Output Fields section allows you to determine the fields that will appear in the attribute table of the resulting feature class.

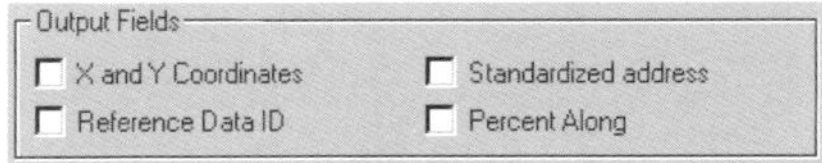

X and Y Coordinates

Use this setting to specify whether or not to create attributes in geocoded feature classes that contain the x and y coordinates of the geocoded features. If you use this setting, then an address locator will create two attribute fields in the output feature class, one for each of the x coordinates and another for each of the y coordinates of the geocoded features. These attributes are not valid for an address if the address is not matched.

Standardized address

This setting allows you to specify whether or not to create an attribute in a geocoded feature class that contains the standardized address. The contents of this field for each address are the address components used by the address locator, separated by the pipe ("|") character. This attribute is useful for determining how the address locator standardized the addresses.

Reference Data ID

Use this setting to specify whether or not to create an attribute in a geocoded feature class that contains the ID of the reference feature to which an address was matched. This attribute is not valid for an address if the address is not matched.

Percent Along

Address locators based on the US One Range, US Streets, or StreetMap address locator styles can interpolate a position along reference features for a geocoded address. Use this setting to specify whether or not to create an attribute in a geocoded feature class that contains the position along the reference feature to which the address was matched. The value of this attribute is a number between 0 and 100, with 0 indicating the starting node of the reference feature and 100 indicating the ending node of the reference feature. This attribute is not valid for an address if the address is not matched.

Creating a new address locator

1. In ArcCatalog, click an address locators folder.
2. Double-click the Create New Address Locator item.
3. Click the address locator style that you want to use to create the new address locator.
4. Click OK. ►

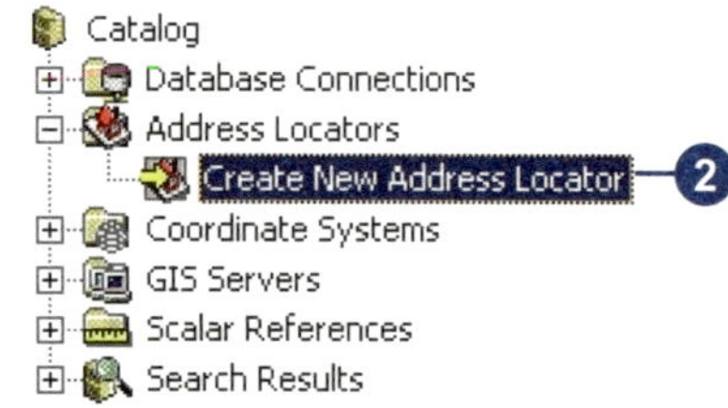

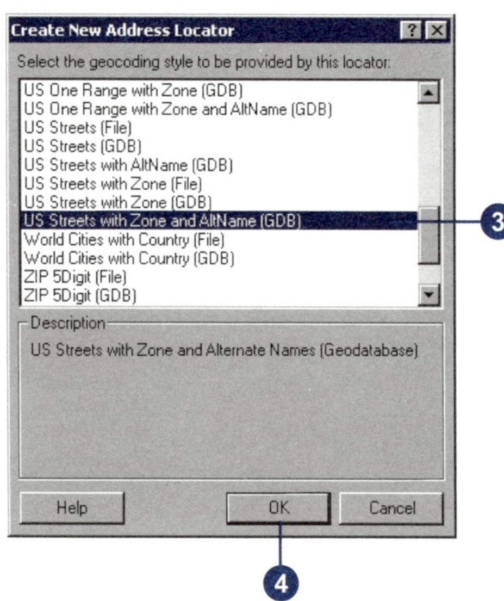

5. In the Name and Description text boxes, type a name and description for the new address locator.
6. Click the Browse button on the Primary table tab.
7. Navigate to and click the feature class or shapefile that the address locator will use as reference data, then click Add.
8. Click a dropdown arrow, then click the name of the field that contains the specified address information.

   The required address elements have a bold field name. ▶

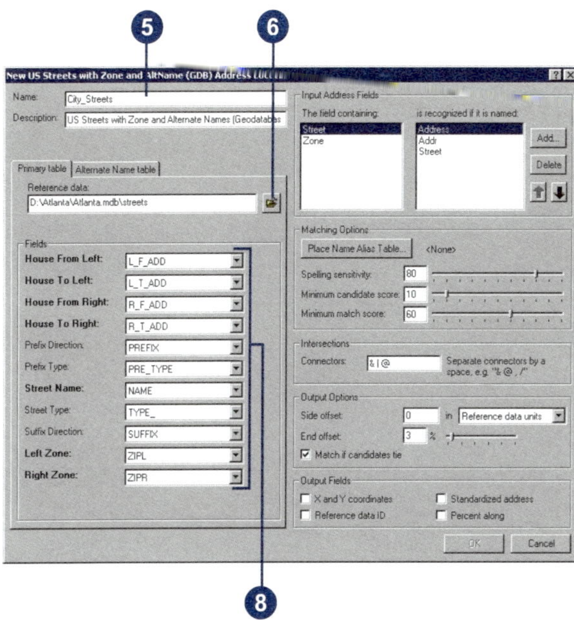

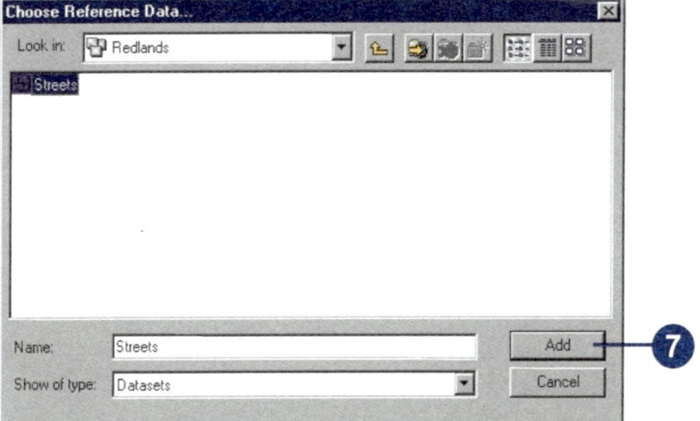

9. If your address locator will use an alternate street name table, click the Alternate Name table tab.

   Otherwise, skip to step 13.

10. Click the Browse button.

11. Navigate to and click the table that the address locator will use as an alternate street name table, then click Add.

12. Click a dropdown arrow, then click the name of the column that contains the specified alternate street name information.

    The required alternate street name attributes have a bold name for the field.

13. If your address locator will use a place name alias table, click Place Name Alias Table.

    Otherwise, skip to step 16. ▸

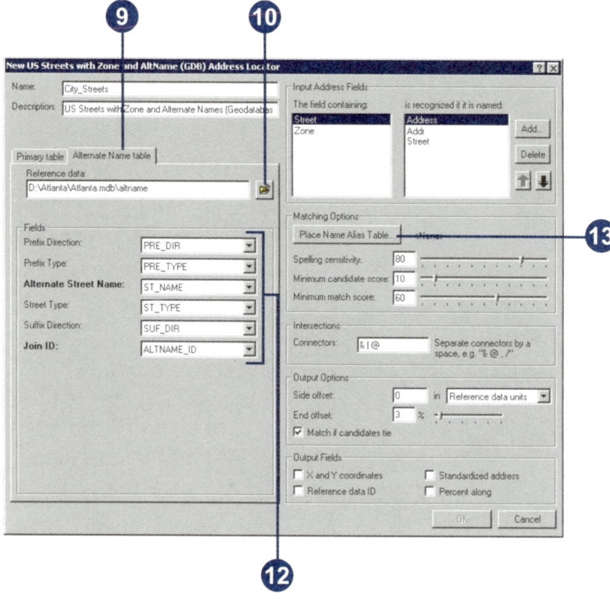

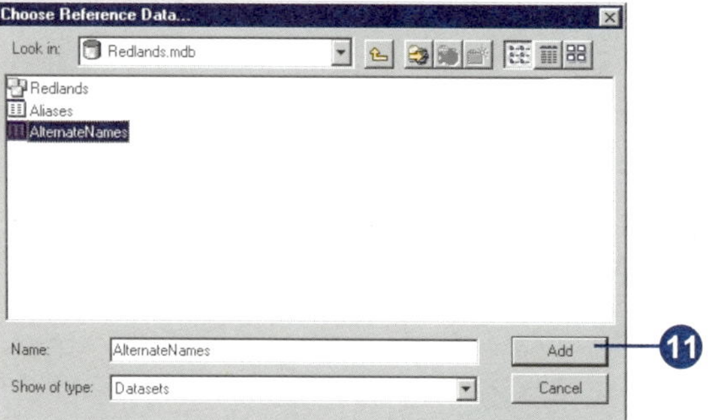

14. Click the Browse button, navigate to and click the table that the address locator will use as a place name alias table, then click Add.
15. Click a dropdown arrow, then click the name of the column that contains the specified place name alias information.

    The required place name alias attributes have a bold name for the field.
16. Click OK.
17. Click OK to create the new address locator.

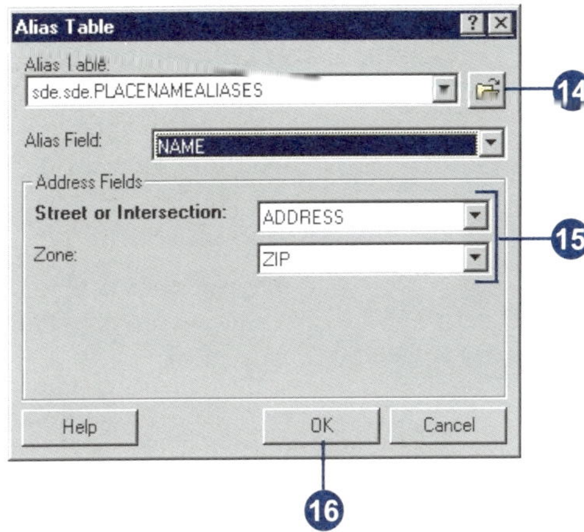

Creating an address locator using ArcToolbox

1. In the ArcToolbox window, open the geocoding toolbox and double-click the Create Address Locator tool.

   This will open the Create Address Locator dialog box.

2. Click the Browse button next to the Address Locator Style text box and navigate to the address locator template file you wish to use in creating the address locator.

3. Click the Browse button next to the Reference Data text box and navigate to the reference data you wish to use in the address locator.

4. Once the reference data is listed in the Reference Data text box, click the Add button next to the Reference data table. This will add the reference data to the address locator.

5. Repeat the process of adding data to the Reference Data table until all reference material is listed.

6. Using the dropdown menu in the Role field of the Reference Data table, indicate the role that the reference material plays in the address locator. ▸

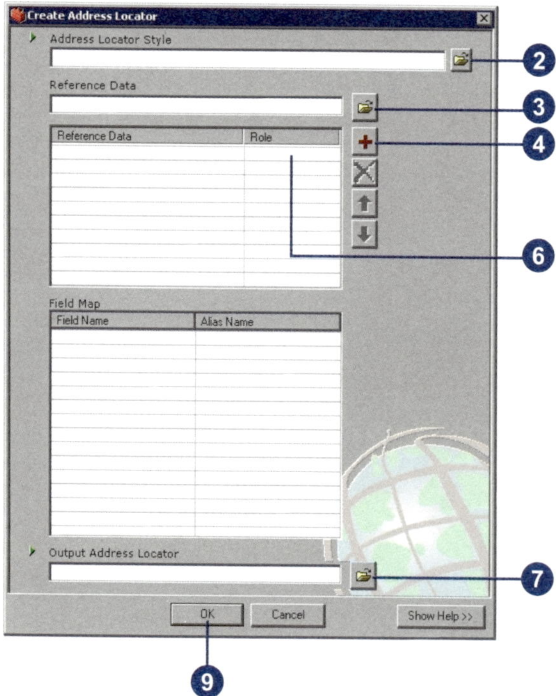

The required fields will appear in the Fields Map portion of the dialog box.

7. Use the Browse button next to the Output Address Locator to determine where you wish to store the address locator.
8. On the Navigate dialog box, specify a name for the address locator and click OK.
9. Click OK on the Create Address Locator dialog box. This will initiate the create address locator process.

Modifying an address locator's settings

You can modify an address locator's settings to control how it determines the locations of addresses and what information is contained in the geocoding output. One option is to set the columns in which you expect to find street names, for example. When the address locator searches an address table for the column containing street names, it looks for columns in the order in which they appear in the list on the right side under Input Address Fields. If the table doesn't contain a column named "Address" it will look for a column named "Addr". You might also set the sensitivity to use when matching addresses, or define which attributes should be added to the resulting feature class.

Tip

Reordering default input column names

To change the order in which the column names appear, click a name, then click the up and down arrows located to the right of the related text boxes in order to move the name up or down in the list.

Adding a default input address column name

1. Right-click the address locator that you want to modify and click Properties.
2. In the list on the left side under Input Address Fields, click the input address column for which you want to search in address tables.
3. Click Add.
4. Type the name of the field to search for in address tables, then click OK.
5. Click OK.

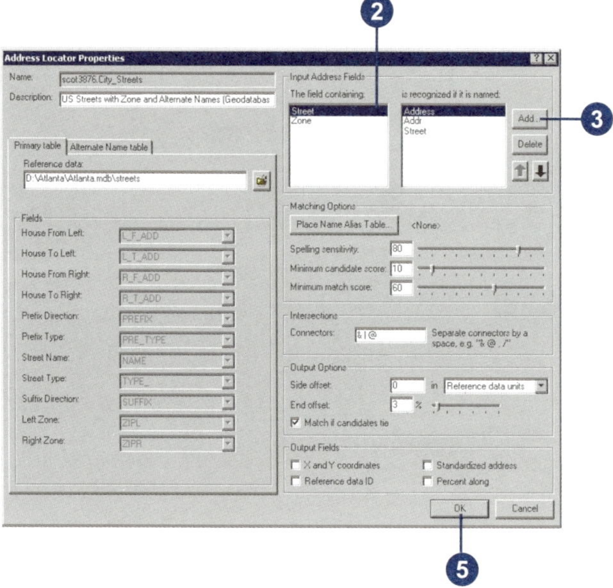

Tip

Specifying an alternate street name table

To set the alternate street name table that will be used by this address locator, see steps 9 through 12 in the task 'Creating a new address locator' in this chapter.

Tip

Specifying a place name alias table

To set the place name alias table that will be used by this address locator, see steps 13 through 15 in the task 'Creating a new address locator' in this chapter.

Modifying matching options

1. Right-click the address locator that you want to modify and click Properties.
2. Under Matching Options, drag the Spelling sensitivity slider to the desired setting.
3. Drag the Minimum candidate score slider to the desired setting.
4. Drag the Minimum match score slider to the desired setting.
5. Click OK.

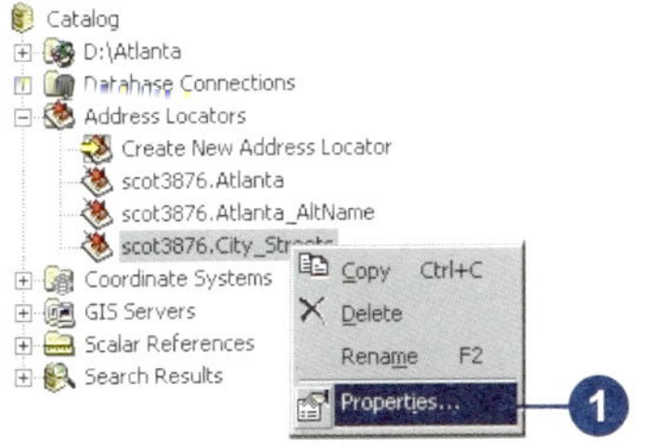

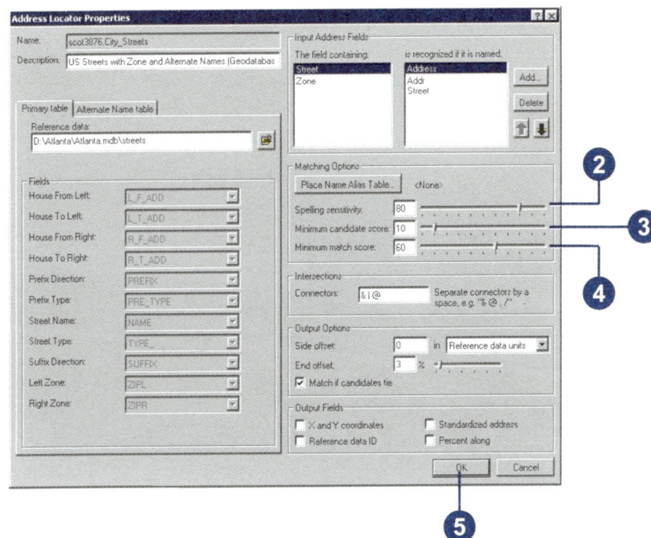

Specifying additional intersection connectors

1. Right-click the address locator that you want to modify and click Properties.
2. Type the intersection connectors that the address locator will recognize in the Intersection Connectors text box.

   Items in the Intersection Connectors text box must be separated by a space (for example, "& @ , |").
3. Click OK.

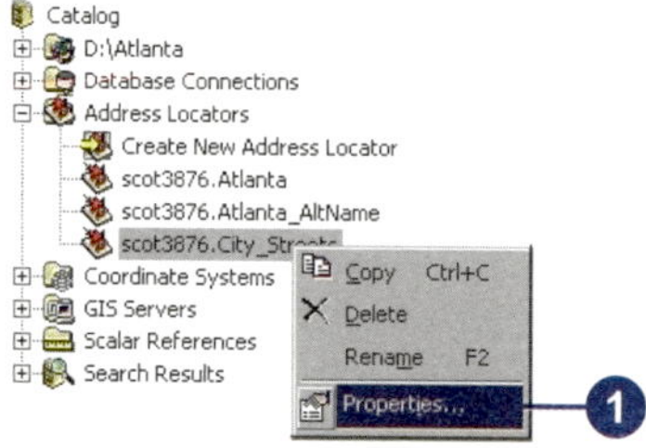

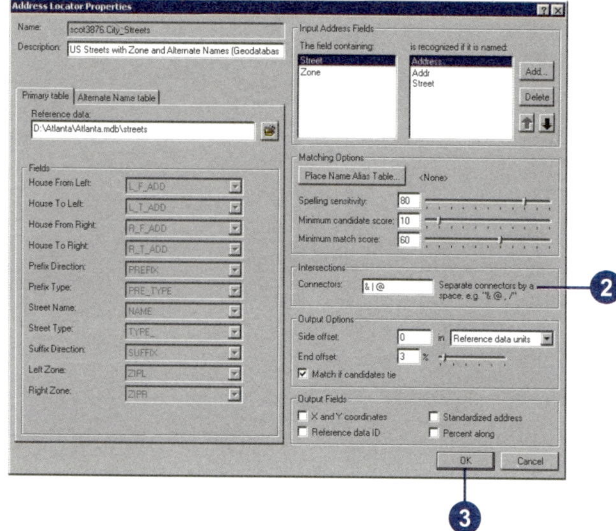

Modifying output options

1. Right-click the address locator that you want to modify and click Properties.
2. Under Output Options, click the dropdown arrow and click the units that will be used to measure the side offset.
3. Type the number of units by which to offset geocoded addresses.
4. Drag the End offset slider bar to the desired setting.
5. Check Match if candidates tie to match addresses arbitrarily when two or more candidates with the same best match score exist.
6. Click OK.

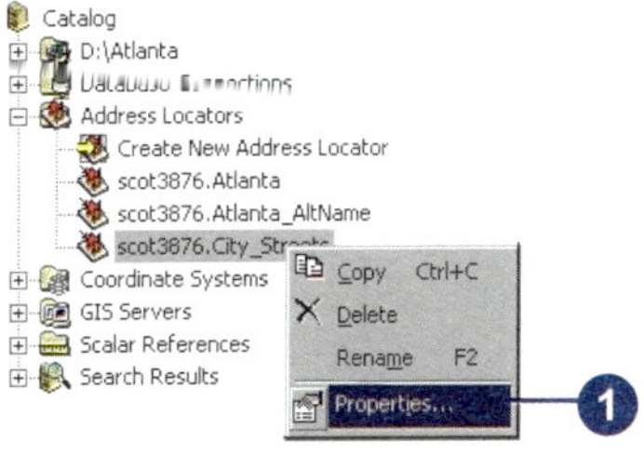

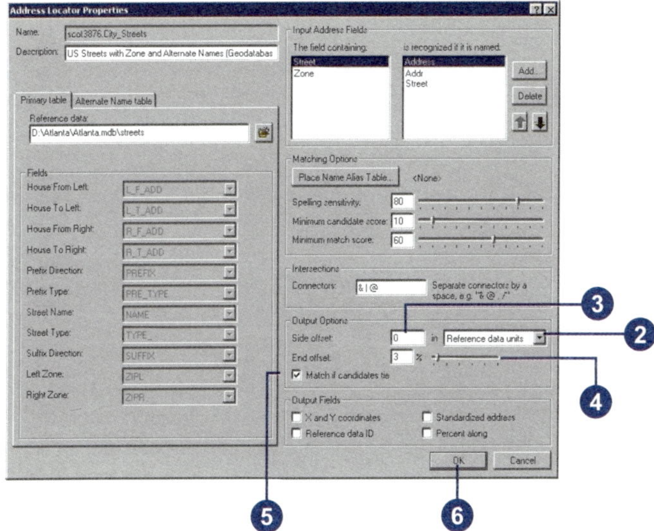

Specifying output field attributes

1. Right-click the address locator that you want to modify and click Properties.
2. Under Output Fields, check X and Y coordinates to write coordinates of geocoded features to geocoded feature classes.
3. Check Standardized address to write standardized addresses to geocoded feature classes.
4. Check Reference data ID to write the identification values of the reference data features to which addresses were matched to geocoded feature classes.
5. Check Percent along to write the percent along reference features at which addresses are located to geocoded feature classes.
6. Click OK.

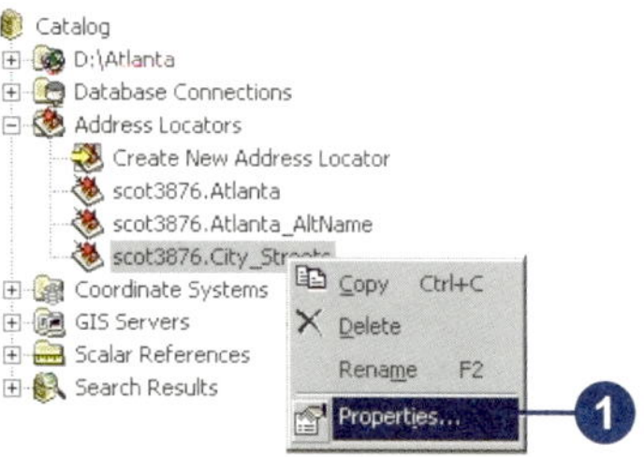

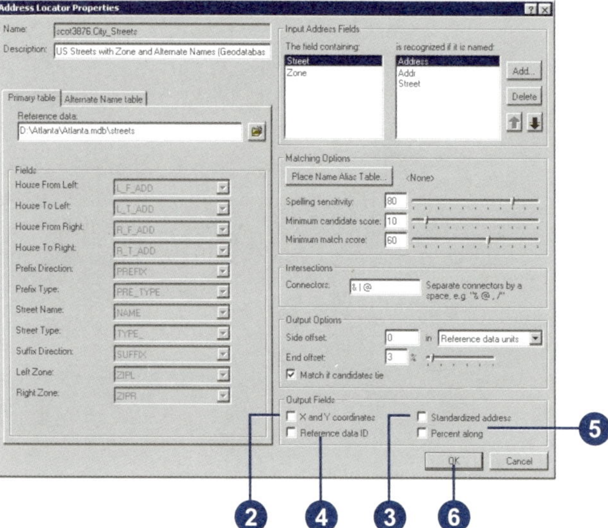

Managing address locators in ArcMap

In ArcMap, you can find addresses or geocode tables of addresses using address locators. Your ArcMap document can contain any number of address locators. You can use the Address Locator Manager to manage the set of address locators contained in an ArcMap document.

Adding an address locator to an ArcMap document

1. Click the Tools menu, point to Geocoding, then click Address Locator Manager.
2. Click Add.
3. Browse to the address locator that you want to add to the ArcMap document and click Add.
4. Click Close.

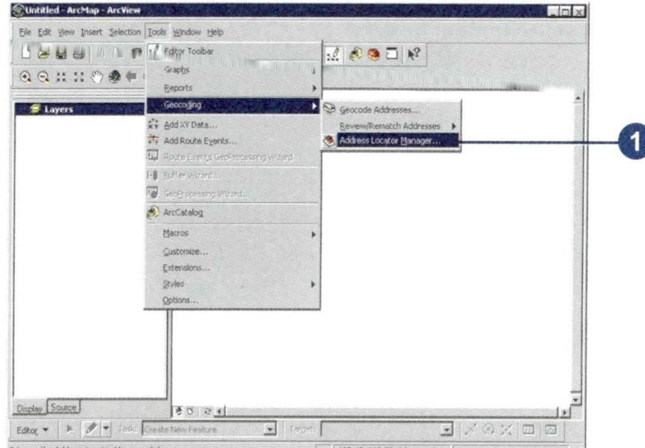

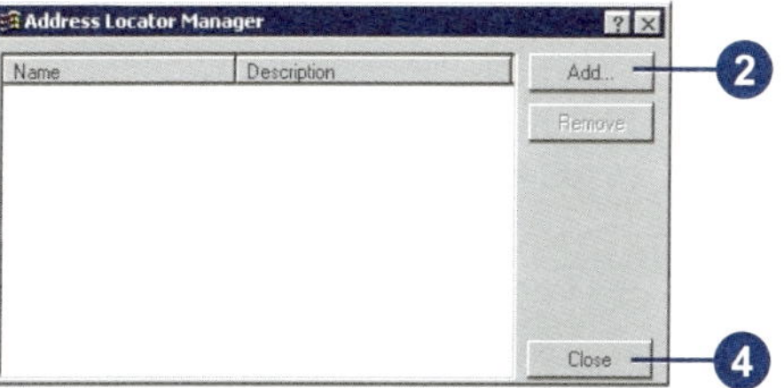

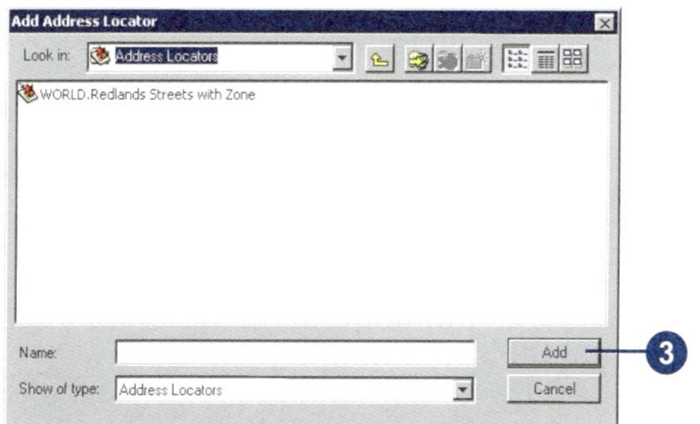

Tip

Saving the address locator

Removing an address locator from an ArcMap document does not eliminate the locator. It simply removes it from that particular ArcMap document.

The address locator is still accessible through ArcCatalog and can be added to other ArcMap documents.

Removing an address locator from an ArcMap document

1. Click the Tools menu, point to Geocoding, then click Address Locator Manager.
2. Click the address locator that you want to remove from the ArcMap document, then click Remove.
3. Click Close.

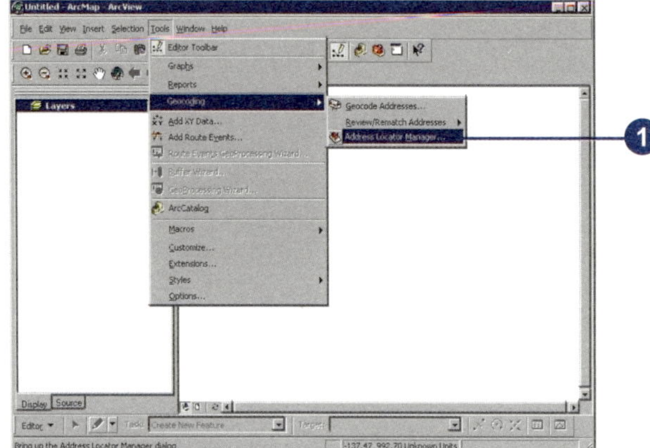

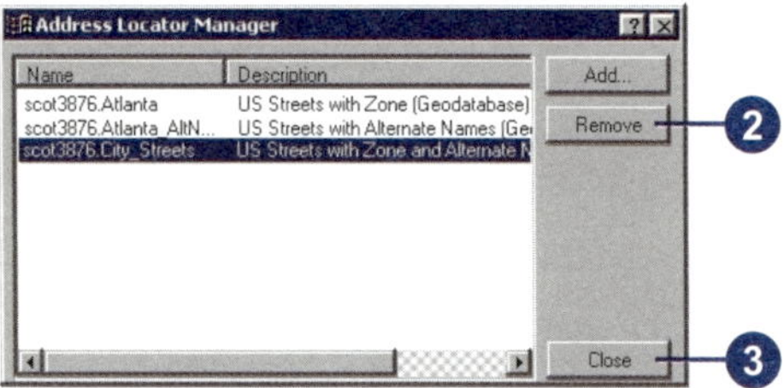

Tip

Deleting an address locator

When you create an address locator, there are many additional index files created. When you no longer need an address locator, you may want to delete the address locator as well as the related index tables. This process is facilitated through a tool available in ArcToolbox.

Deleting an address locator in ArcToolbox

1. In ArcToolbox, open the geocoding toolbox and double-click the Delete Address Locator tool.

   This will open the Delete Address Locator dialog box.

2. Click the Browse button next to the Address Locator text box and navigate to the address locator you wish to delete.

3. Click OK on the Delete Address Locator dialog box.

   This will initiate the deletion process.

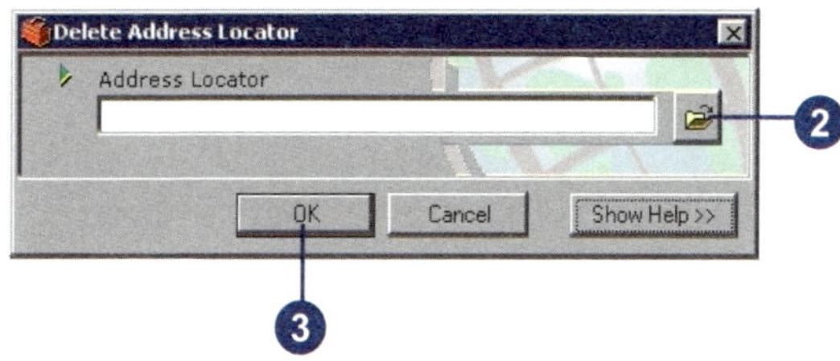

Locating addresses

6

IN THIS CHAPTER

- **Understanding the geocoding process**
- **Finding an individual address**
- **Geocoding a table of addresses**
- **Rematching a geocoded feature class**

The central goal of geocoding is to find the locations of addresses. Once you have prepared your address and reference data and created your address locator, the process of finding an address is relatively easy.

This chapter begins with a detailed description of the geocoding process including how your address data is interpreted, translated, and searched against the reference data. Knowing this process will help you understand the intricacies of geocoding and will help you determine what the geocoding process requires in terms of address data. Several tasks are also outlined to guide you through the process of finding a location and rematching addresses.

Understanding the geocoding process

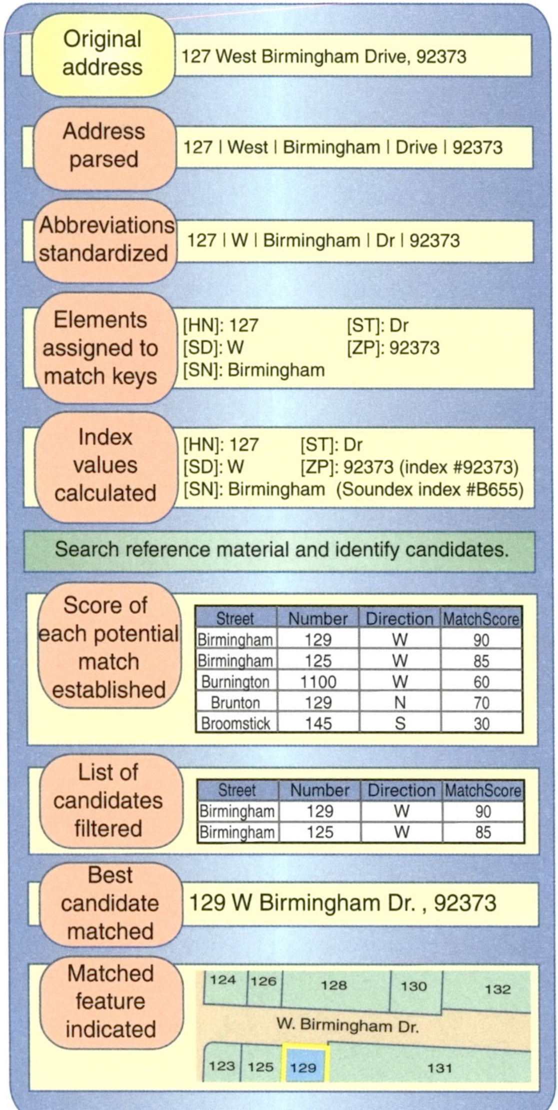

Up to this point, you have learned many of the steps in preparing for geocoding. With this done, you can begin to geocode addresses. In this chapter you will learn how an address locator prepares the address data, searches the reference data, and matches addresses. You will also learn how modifying an address locator's settings affects this process and can help you improve the accuracy of your geocoding search.

Address locators use a specific process to find a location. First, the search parameter—often an address—is parsed into individual address components. Next, the address locator standardizes the address, defining correct format of presentation. The address elements are then assigned to specific categories used in the search. The address locator then searches the reference data to find potential candidates. Each potential candidate is assigned a score based on how closely it matches the address. Finally, the address is matched to the candidate with the best score.

Address parsed

When an address locator parses an address, it dissects the address into its distinct address components. For example, the address "127 West Birmingham Drive, 92373" has five address components: the street number, "127"; the street name, "Birmingham"; the street type, "Drive"; the street direction, "West"; and the postal code, "92373". Each style of address locator parses an address into a different set of address components.

When geocoding using place name alias tables, the address locator recognizes the input address as a place name rather than an address. The place name alias table is searched, a match is found, and the primary attribute table is referenced. From there, the process continues as outlined. The intersection of two streets is also recognized by the address locator as an input address and is searched using solely the street attributes. More information on geocoding using these alternative approaches is presented in the chapter 'Additional geocoding techniques' in this book.

Abbreviations standardized

Many elements of an address, such as direction or street type, are often written using an abbreviation. For most of these address elements, there is a wide range of formats that are used to present them. For example, the direction of west could be written using "West", "W.", or "Wst." The term "drive" could also be written using "Dr.", "Drv.", or "Drive". Attempting to geocode using such a wide range of abbreviations would greatly increase the search time. To avoid this, the address elements that are often abbreviated are assigned a standardized value. This allows the search to include only the standardized value, thus reducing the search time required.

Address elements assigned to match keys

Each individual element in the address being searched is assigned to a particular category. These categories are referred to as match keys. When your address locator was created, corresponding categories were established for the material in your reference data. Match keys are used to compare the address elements to attributes that were defined in the reference data.

Match keys have a two-letter coded value and include categories such as house number, assigned the coded value of HN; street direction as SD; street name as SN; street type as ST; ZIP Codes as ZP; and many others.

Index values calculated

As discussed previously in this book, when creating an address locator, geocoding indexes are produced for the reference data. This speeds the process of searching the potentially large list of entries in the reference data. When searching for an address, corresponding elements of the address are assigned index values that match those created in the geocoding index. These index values are used to quickly match the address attributes to fields in the geocoding index.

Reference data searched

Once the address locator has standardized the address, it compares the elements of the address data to those in the reference data to find features with address elements that are similar to the components of the standardized address. Each style of address locator bases this search on a different set of address elements. The address locator uses its spelling sensitivity setting for some address elements, such as street name, to determine how closely the address elements of a feature must match the address elements of the address you are geocoding. If the address locator uses an alternate street name table, then it also searches this table to find potential candidates.

Score of each potential match established

The geocoding engine assigns a score to each candidate in order to determine how closely each one matches the address that you are geocoding. Each potential candidate is assigned a score from 0 to 100. Each address element is used to generate this score. The score for each potential candidate will be lower if address elements are misspelled (for example, the street name is misspelled), incorrect (for example, the street number of the address does not fall within the address range for the candidate), or missing (for example, if the street direction is specified in the reference data but not in the potential candidate).

List of candidates filtered

With a score established, the address locator generates a set of candidates that are potential matches for the address. Those potential matches that have a score lower than the minimum candidate score are removed from the potential match list. For more information on the minimum candidate score, see the chapter 'Preparing for geocoding' in this book.

Best candidate matched

Finally, the address locator finds the candidates with the highest score. If the score of the candidate with the best score exceeds the address locator's minimum match score setting, then the address locator matches the address to that candidate. Otherwise, the candidate list is presented, allowing you to select the best match.

Finding an individual address

You can use address locators to find addresses in ArcMap. In order to use an address locator in ArcMap, it must be loaded into the ArcMap document. You don't need to load the reference data for an address locator into the ArcMap document, but doing so will give you a visual reference to help you choose an appropriate candidate for an address.

Tip

Adjusting geocoding settings

If your search results in too many or too few candidates, you can modify the geocoding settings that the address locator uses to find addresses. Modifying the geocoding settings in ArcMap does not change the address locator that you are using. Only the settings that are used to find addresses in your ArcMap session are modified.

1. In ArcMap, click the Find tool.
2. Click the Addresses tab.
3. Choose the address locator that you want to use to find the address from the list.

   If the address locator that you want to use does not appear in the list, you need to add it to the ArcMap document. Click the Browse button to browse for address locators to add to the document.
4. Type the address components in the text boxes.
5. Click Find.
6. Click Show all candidates to see all of the candidates that the address locator generated.

   By default, only the candidates that meet or exceed the minimum match score are shown.
7. If you want to modify the geocoding settings that the address locator uses to find the address, click Options.

   Otherwise, skip to step 9. ►

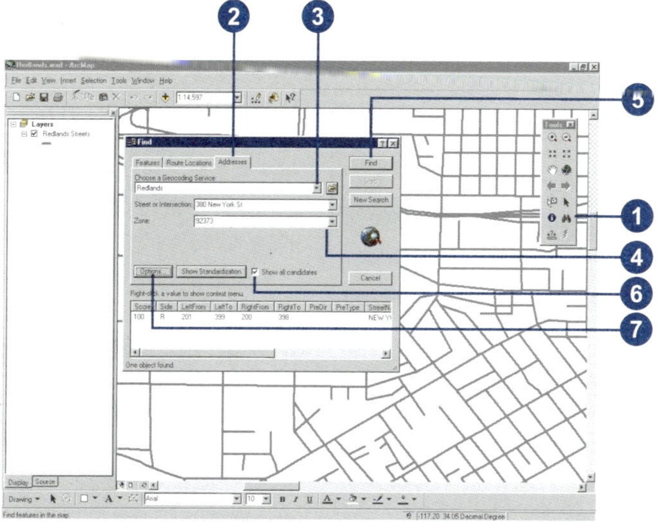

See Also

For more information on the geocoding options, see the chapter 'Preparing for geocoding' in this book.

Tip

The Edit Standardization dialog box

If your search returns no candidate or the wrong candidate, you may find that editing the standardized address will improve your results.

8. Modify the address locator's settings and click OK.
9. If you want to change how the address locator standardized the address, click Show Standardization.

   Otherwise, skip to step 11.
10. Edit the address's standardization by editing the values in the fields in the Edit Standardization dialog box.

    When you make edits in the Edit Standardization dialog box, the list of candidates is automatically updated. ►

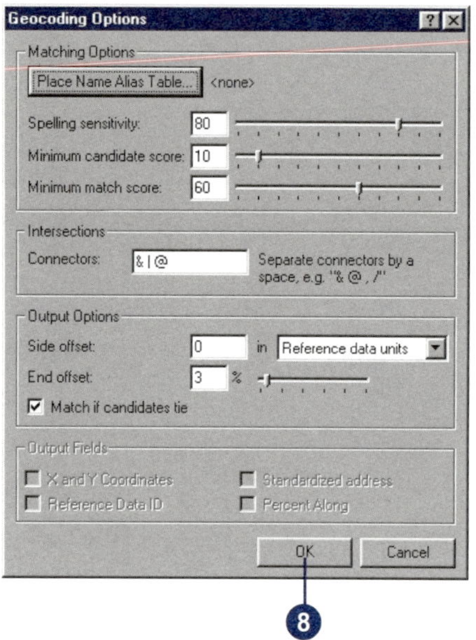

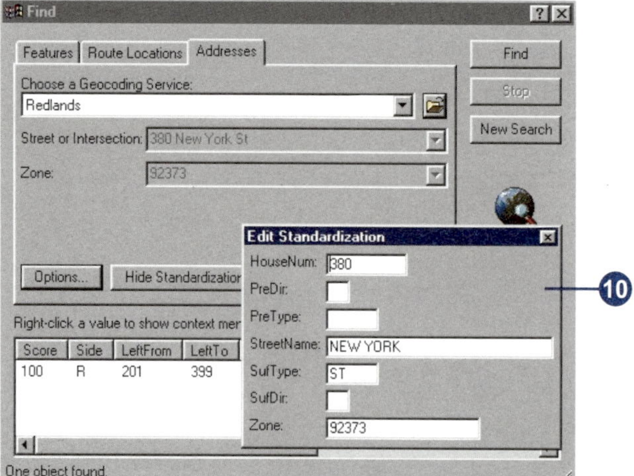

11. Right-click the candidate and click Flash Candidate Location(s) to flash the location of a candidate.
12. Right-click the candidate and click Zoom to Candidate(s) and Flash to zoom in to a candidate.
13. Right-click the candidate and click Add as Graphic(s) to Map to add a graphic to the map at a candidate's location.
14. Right-click the candidate and click Set Bookmark to set a spatial bookmark for a candidate's location.
15. Click Cancel to close the Find dialog box.

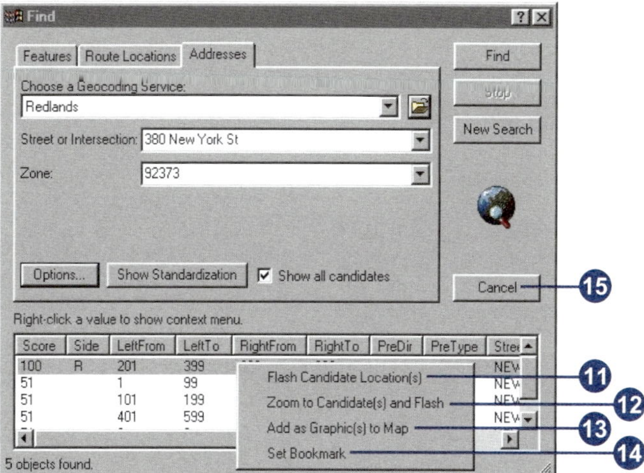

Geocoding a table of addresses

When you geocode a table of addresses, you use an address locator to create point features that represent the locations of the addresses. You can geocode a table of addresses into any spatial format supported by ArcGIS including ArcSDE feature classes, personal geodatabase feature classes, and shapefiles. ArcView users cannot create ArcSDE feature classes.

Before you can geocode a table of addresses, you must create an address locator and prepare your table to be geocoded. For information on preparing your address tables for geocoding, see the chapter 'Preparing for geocoding' in this book.

Tip

Geocoding a table of addresses in ArcCatalog

In ArcCatalog, right-click the table of addresses you wish to geocode and select Geocode Addresses. This will initiate the same series of dialog boxes that are present in ArcMap.

1. In ArcMap, click the Tools menu, point to Geocoding, then click Geocoding Addresses.

2. Click the address locator that you want to use to geocode the table of addresses and click OK.

   This will close this dialog box and open the Geocode Addresses dialog box.

   If the address locator that you want to use does not appear in the list, click Add to browse for the address locator. ▶

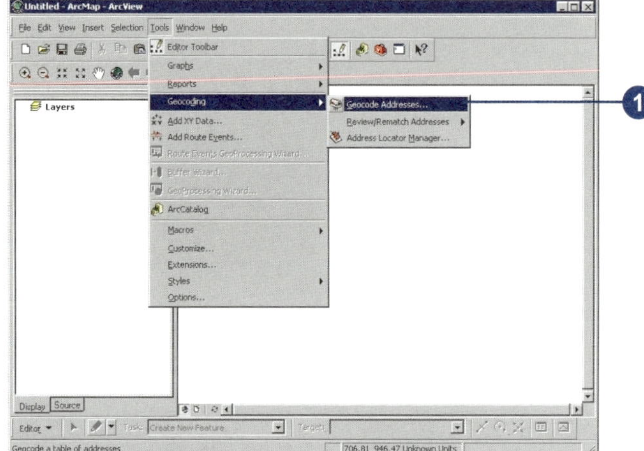

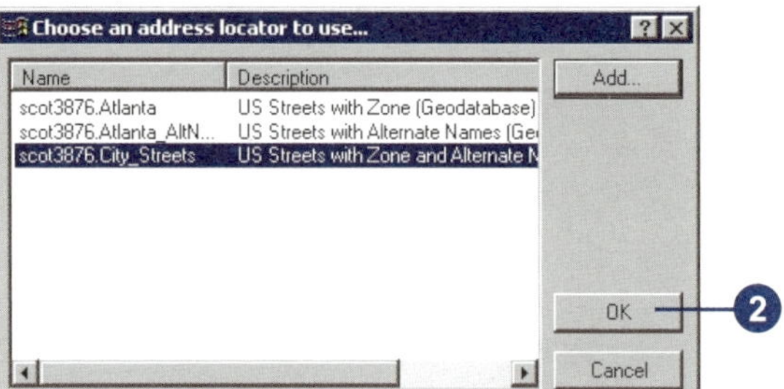

Tip

Geocoding dynamic feature classes related to the address table

A dynamic feature class is one that will be updated whenever the material on the related address table is altered.

If you want to create a dynamic feature class related to the address table, the address table and geocoded feature class must be in the same geodatabase. This option is not available to ArcView users.

3. Click the dropdown arrow and click the table that you want to geocode.

   If the table does not appear in the dropdown list, click the Browse button to browse for the table.

4. Choose the column name from each dropdown list that contains the specified address information.

   The names of the required address attributes are shown in bold.

5. Click the Browse button and navigate to the location where you want to create the geocoded feature class.

6. Click Save.

7. To create a geocoded feature class that has a dynamic relationship with the address table, click the option to do so.

8. Click Advanced Geometry Options to specify the geometry settings for the geocoded feature class. ►

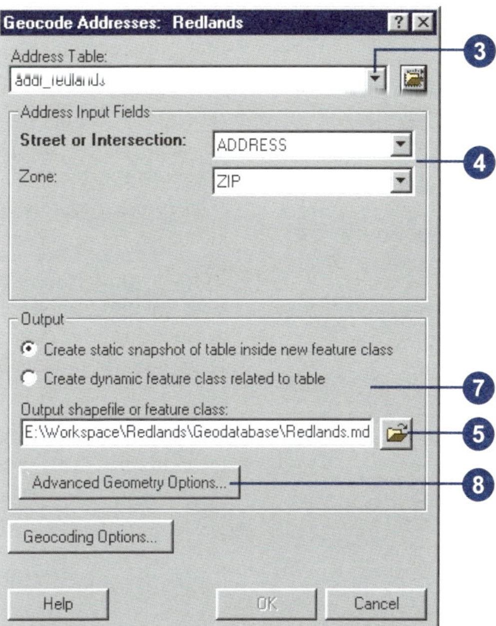

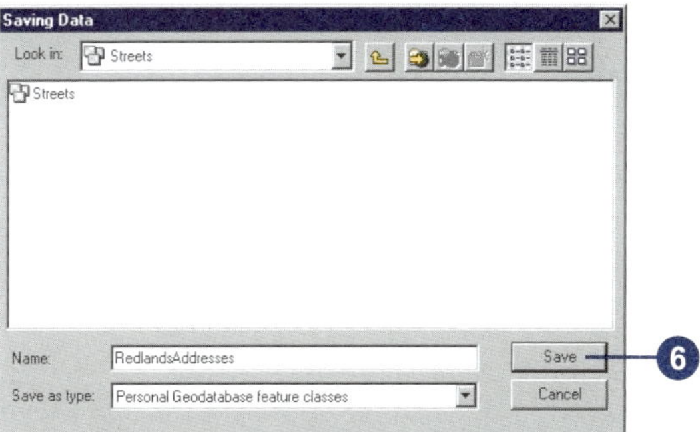

Tip

Specifying the geocoding settings

When you are geocoding a table of addresses, modifying the geocoding settings does not change the address locator that you are using. Only the settings that are used to geocode this table are modified. These settings are stored with the geocoded feature class. The original address locator is not modified.

For more information on modifying an address locator's settings, see the chapter 'Preparing for geocoding' in this book .

9. Specify the geometry settings for the geocoded feature class and click OK.
10. To specify the geocoding options that will be used to geocode the table of addresses, click Geocoding Options.
11. Specify the geocoding settings that you want to use to geocode the table of addresses and click OK.
12. Click OK on the Geocode Addresses dialog box to geocode the table. ▶

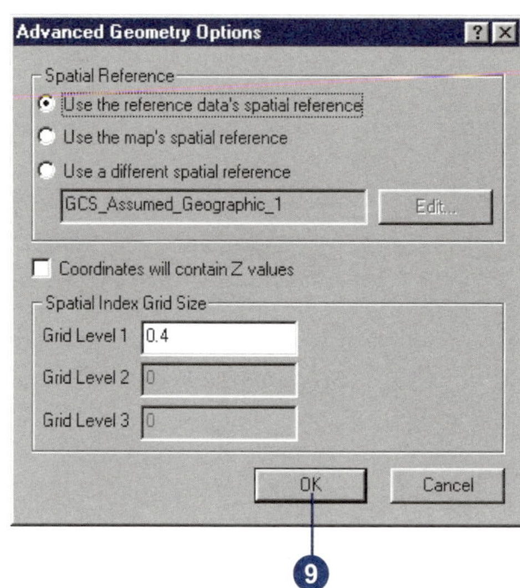

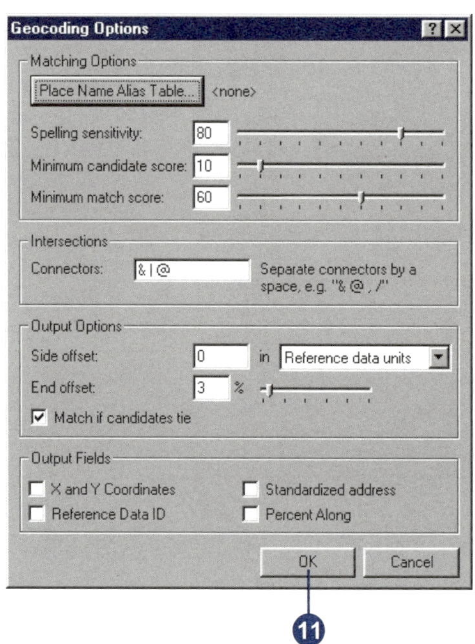

Tip

Attributes in geocoded feature classes

When you geocode a table of addresses, an address locator creates some special attributes in the output feature class.

The Status attribute indicates whether or not the address was matched. This attribute has values of "M" for matched addresses, "U" for unmatched addresses, and "T" (tied) for addresses for which there were more than one candidate with the best match score.

The Score attribute contains the match score of the candidate to which the address was matched.

The Side attribute contains the side of the street to which an address was matched, if the address locator that was used to match the table contains address information for both sides of the street. This attribute has values of "L" for the left side of the street, "R" for the right side of the street, or nothing if the address locator could not determine the side of the street.

13. Review the results of the geocoding process.
14. Click Done.

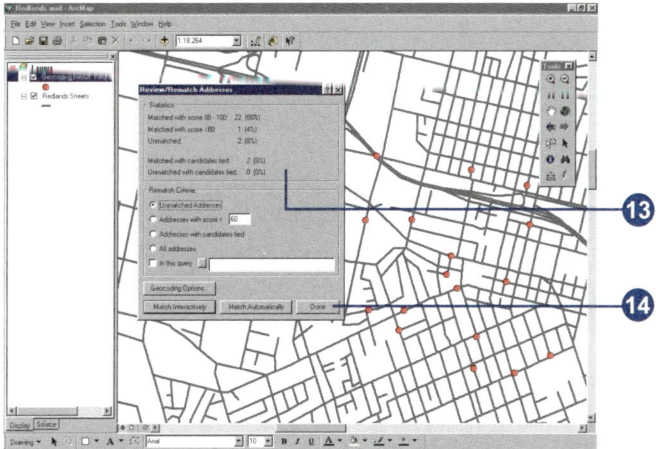

Geocoding a table of addresses in ArcToolbox

1. In ArcToolbox, open the geocoding toolbox and double-click the Geocode Addresses tool.

   This will open the Geocode Addresses dialog box.

2. Click the Browse button next to the Input Table text box and navigate to the address table you wish to geocode.

3. Click the Browse button next to the Address Locator text box and navigate to the address locator you wish to use to geocode the address table.

   The input address field names and their aliases will appear in the Input Address Fields table.

4. Click the Browse button next to the Output Feature Class text box and navigate to the folder or geodatabase where you wish to store the output address table.

5. In the Navigate dialog box, specify a name and file type for the output address table and click OK. ▸

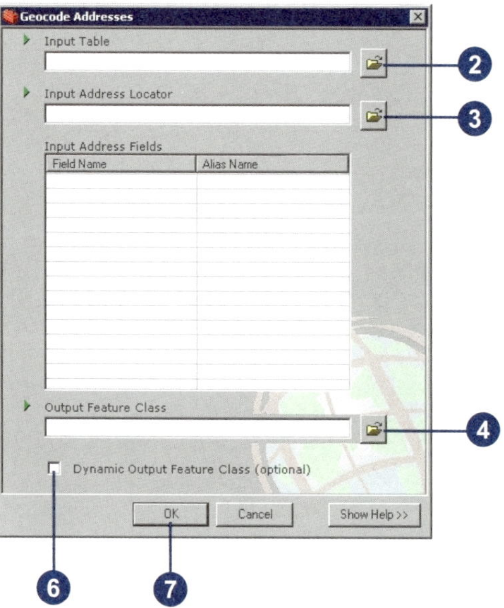

6. If you wish to create a dynamic feature class, click the check box to do so.
7. Click OK on the Geocode Addresses dialog box to initiate the geocoding process.

Rematching a geocoded feature class

After you have geocoded a table of addresses, you may want to review the results. If you are unhappy with the results, you may want to modify the address locator's settings and try to geocode the table of addresses again. This process is known as *rematching*.

There are a number of options for specifying which addresses in a geocoded feature class you want to rematch. You can rematch just the addresses that are unmatched, all of the addresses with a match score less than a certain value, all of the addresses with two or more candidates with the best match score, or all of the addresses. In addition, you can specify a query to use that defines the set of addresses to rematch. For more information on defining queries, see the book *Using ArcMap*.

Tip

Starting an edit session

When beginning a review/rematch session, you may be prompted to start an edit session. By clicking yes, you are beginning to modify the feature class created through the geocoding process.

Rematching a geocoded feature class automatically

1. In ArcMap in the table of contents, right-click the feature class that you want to rematch, click Data, then click Review/Rematch Addresses.

   This will open the Review/Rematch Addresses dialog box.

2. Specify the criteria for the addresses that you want to rematch.

3. Click Geocoding Options to modify the geocoding settings that you want to use to rematch the addresses. ►

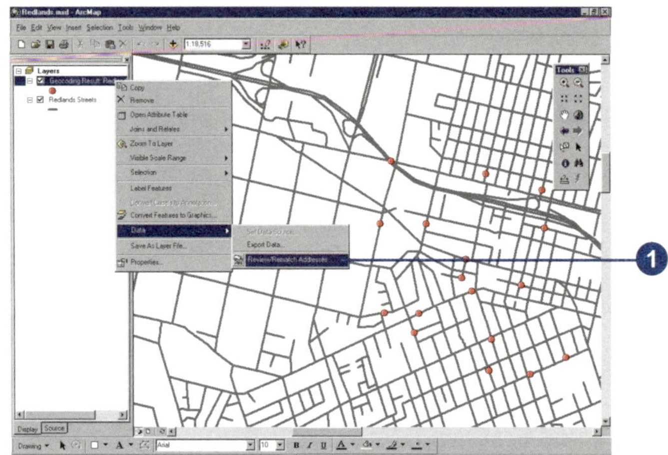

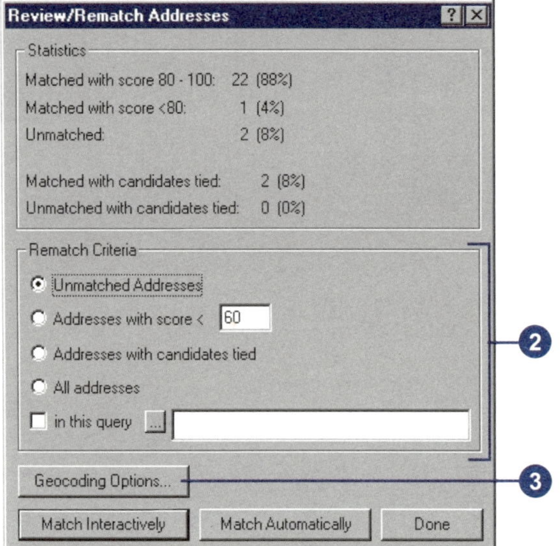

Tip

Specifying the geocoding settings

When you are rematching a geocoded feature class, modifying the geocoding settings does not change the address locator that was originally used to create the feature class. The geocoding settings that are presented here are the settings that were used to geocode the feature class.

For more information on modifying an address locator's settings, see the chapter 'Preparing for geocoding' in this book .

Tip

Automatically rematching addresses in ArcCatalog

In ArcCatalog, right-click the feature class that you want to rematch and click Review/Rematch Addresses. This will initiate the same series of dialog boxes that are present in ArcMap.

4. Specify the geocoding settings that you want to use to rematch the geocoded feature class and click OK.
5. Click Match Automatically.
6. Review the results of rematching the specified addresses.
7. Click Done.

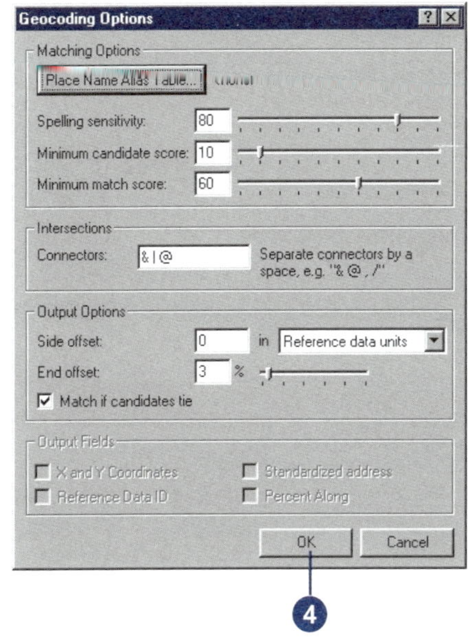

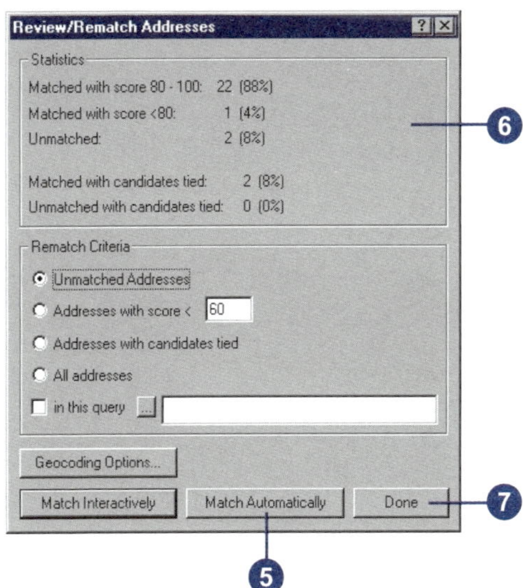

Tip

Interactively rematching addresses in ArcCatalog

In ArcCatalog, right-click the feature class that you want to rematch and click Review/Rematch Addresses. This will initiate the same series of dialog boxes that are present in ArcMap. However, ArcCatalog does not provide the map interface used in confirming the address location.

Rematching a geocoded feature class interactively

1. In ArcMap on the table of contents, right-click the feature class that you want to rematch, click Data, then click Review/Rematch Addresses.

   This will open the Review/Rematch Addresses dialog box.

2. Specify the criteria for the addresses that you want to rematch.

3. Click Match Interactively. ►

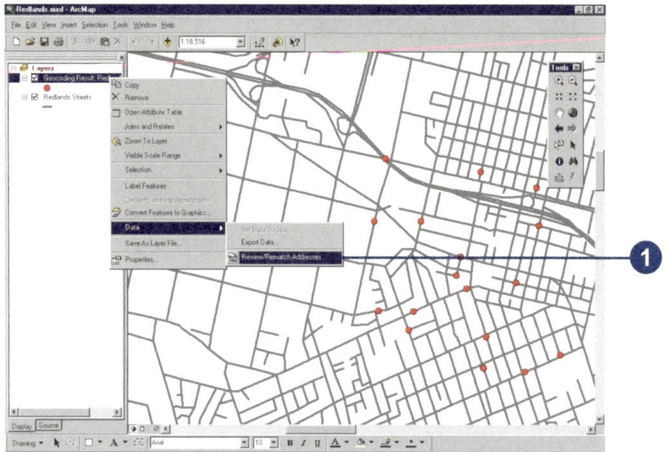

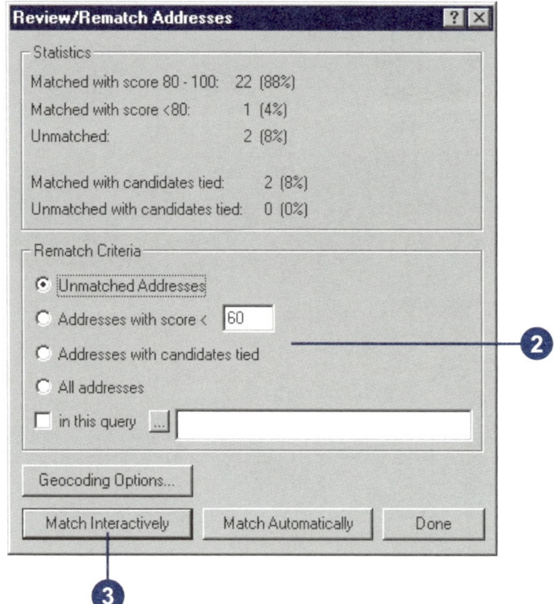

4. Click the address in the top list that you want to rematch.

5. If necessary, edit the input address.

6. Check the address standardization to ensure that the address locator has standardized the address correctly. If not, click Modify to edit the address standardization.

7. Click Geocoding Options to modify the geocoding settings that you want to use to geocode the addresses.

8. Specify the geocoding settings that you want to use to geocode the addresses and click OK.

9. Click Search to refresh the list of candidates.

   The candidates are highlighted on the map.

10. Click Candidates to zoom to the set of candidates for the address.

11. Click the candidate in the bottom list to which you want to match the address.

    The candidate that you select in the bottom list is highlighted on the map in a different color.

12. Click Match. ►

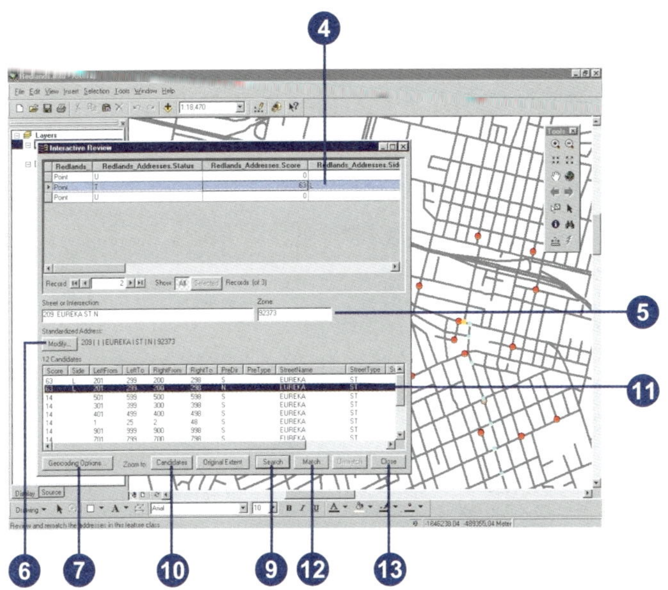

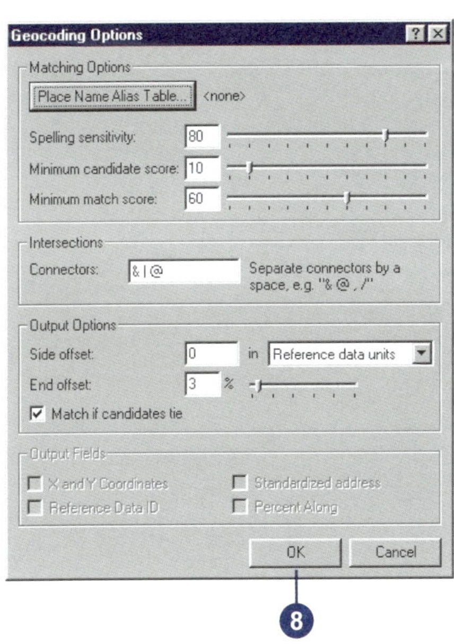

Tip

Unmatching an address

Sometimes you may want to unmatch an address. For example, the address locator may have matched an address to a candidate, but you may not be happy with any of the candidates for the address. Click Unmatch on the Interactive Review dialog box to unmatch an address.

13. Click Close when you are finished rematching the addresses.
14. Click Done.

Additional geocoding techniques

7

IN THIS CHAPTER

- **Alternative searches**
- **The Alternate Name table tab**
- **The Alias Table dialog box**
- **Building an address locator for alternative searches**
- **Searching for an intersection**

Up to this point, this book has discussed the geocoding work flow in terms of searching for locations based on traditional address formats. At times, however, you may find the need to search for locations based on attributes that fall outside of the standard address format. For example, you may need to search for a particular zone such as a postal code or census tract. You may wish to perform a search based on the name of a location, such as a shopping center or business office. Some descriptive elements about a location may have a variety of forms, such as a road referred to by a particular highway number as well as a street name. You may want to create a composite address locator that will first search for a house number, and if no candidates result, will search based on city or postal code information. You may also have data more closely related to an intersection rather than an address such as the location of an auto accident.

Each of these additional geocoding techniques can be done in ArcGIS. In building the address locators for these, you will need to interact with other dialog boxes and perform additional steps. This chapter will describe these alternative methods and the additional steps needed to perform these types of geocoding tasks.

Alternative searches

Searching for nonaddress-defined entities

Depending on your particular application, you may find that the traditional address format required by many of the address locator styles does not accommodate the type of search that you are needing to perform. One example of this is when looking for specific regions or areas. For example, you may need to geocode values according to census tracts or county boundaries. Your application might require that you geocode values to particular watersheds or bioregions. Searching based on properties of a specific region can be performed using the Single Field address locator.

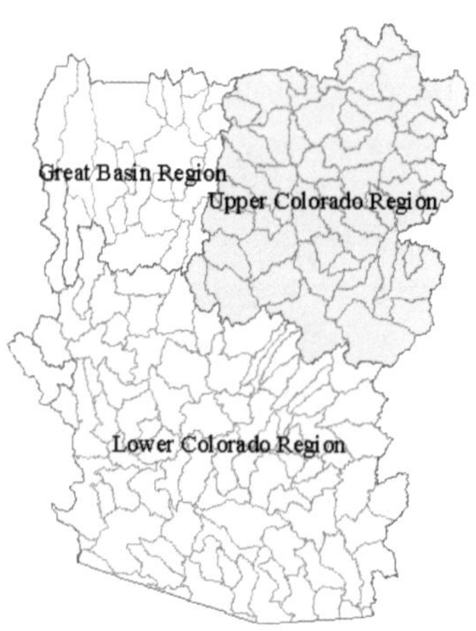

In a manner similar to how you would create an address locator to search for a city or state, you can build an address locator to search for a region or area. Just as in creating other address locators, you must begin with the reference data that contains the entities for which you wish to search. In the example of geocoding watersheds, you would need a watershed feature class. When creating the address locator, you would select the key field on the Primary table tab indicating the field, such as watershed name, in the reference data that you wish to use as the search criteria. When performing the search, you would then indicate this attribute that was selected as the key field. The geocoding engine would process the search parameters against the reference data, delivering an x and y coordinate value for the centroid of the entity.

Address properties with alternative names

At times, streets or cities in your address data are referred to by more than one name. For example, a highway might be known by a name as well as by a particular highway number. Street names also change over time. Further, at times, the names of cities are changed or referred to by another name. In these instances, you may find that your address data refers to the same location using a variety of alternative names.

In order to geocode these locations, ArcGIS provides a method of introducing an alternate name table. Once the address locator has been created with alternate names, you can geocode locations based on the name provided on the primary data table or the alternative name. Simply enter the alternative name into the input address, and the alternate name table will automatically be searched. More information on building an address locator to use alternative names is presented later in this chapter.

Searching for place names

Your application might require the use of location names, such as schools or hospitals, to be used as your input address.

This is feasible if you provide the address locator with a table containing the location or alias name as well as the address data required by your address locator style.

Once the address locator has been created and provided with an alias table, you can geocode locations based on the alias or location name. This is done by entering the alias or location name into the search parameter. The alias table will then be searched. More information on building an address locator to use alias names is presented later in this chapter.

Finding an intersection

At times, you may find that instead of indicating the house number and street, you need to geocode the street intersections. Each address locator style that accommodates street address data has also been provided a means of translating the names of two streets into x and y coordinate values. More information on performing such a search is provided in the section 'Searching for an intersection' later in this chapter.

The Alternate Name table tab

The Alternate Name table tab is located on the Address Locator Properties dialog box next to the Primary table tab. This is the interface used to indicate to the address locator that a table containing alternative names will be used in the geocoding process. This tab is only present when an address locator style that contains reference to an AltName is selected. There are several elements on this tab that are used to direct the address locator.

The Alternate Name table tab is the interface used to specify the table to be used in referencing address elements that have multiple names.

Reference data

The Reference data text box and Browse button allow you to specify the location and name of the Alternate Name table. Upon clicking the Browse button, a navigation interface, much like those used by other applications, will be opened, allowing you to navigate to the desired folder and file. The Alternate Name table reference data acts as another component of the address locator reference data. The required fields in the Alternate Name table will vary based on the address locator style. In addition to the style-specific requirements, each Alternate Name table must contain a column containing the alternative name as well as a column containing a Join ID value, referencing the entity back to the primary table object ID field. For more information regarding reference data requirements for the variety of address locator styles, see the chapter 'Preparing for geocoding' in this book.

A geocoding index is also created for this table; editing or updating this table will require updating the related geocoding index. For more information on geocoding indexes, see the chapter 'Keeping an address locator current' in this book.

Fields

The Fields portion of this tab refers to the columns or fields of the reference data table. The particular field names present reflect the type of address locator style being used and will vary accordingly. Some of the field names are labeled on this interface in bold type. These always include the Alternate Street Name and the Join ID columns. The bold typeface indicates that these fields are required. The other fields labeled with the regular typeface can also be matched to return more accurate results in the location search.

The Alias Table dialog box

The Alias Table dialog box allows you to indicate to the address locator that place names will be used as the search parameter. This dialog box references a table that contains a list of place names and their associated address information. When a place name is used as a search parameter, the address locator searches for the place name on this table, then searches for the location based on the address information associated with it. The Place Name Alias Table acts as another component of the reference data.

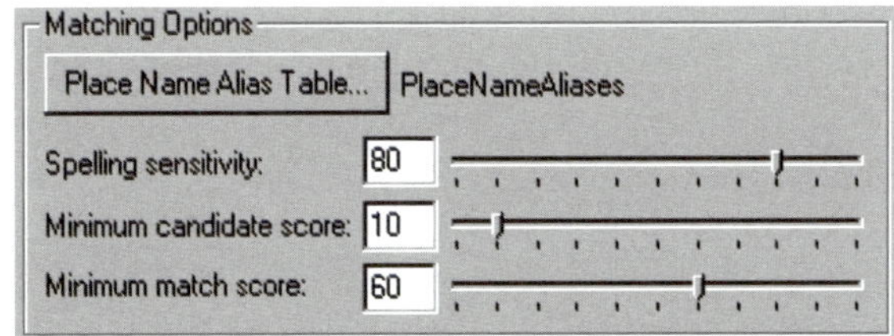

The Alias Table dialog box is opened by clicking the Place Name Alias Table button on the Matching Options portion of the Address Locator Properties dialog box.

Alias Table

The Alias Table text box and Browse button allow you to specify the location and name of the alias name table. Upon clicking the Browse button, a navigation interface much like those used by other applications will be opened, allowing you to navigate to the desired folder and file.

The Alias Table has some required fields. The first of these is the Alias Field. The other required and optional fields are based on the style of address locator you are creating.

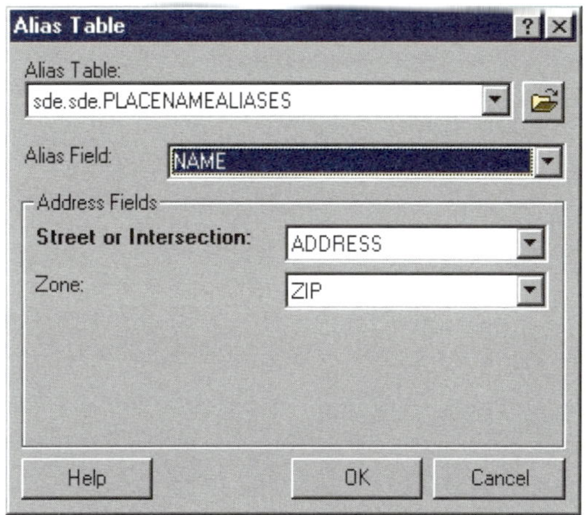

The Alias Table dialog box refers the address locator to a table that contains location names and their addresses. With this, a location name can be used as the search parameter.

Alias Field

The Alias Field refers to the column or field of the Alias Table containing the place name that will potentially be entered as the input address. For example, if the Alias Table contains a list of schools with their associated addresses, the column in the table that contains the actual school name will be the Alias Field. This indicates to the address locator the column in the table to be used in the event that a place name is given as an input address.

Address Fields

The Address Fields portion of this dialog box refers to the columns of fields of the Alias Table and are the same as the set of address input fields used by the address locator. Several of the field names are labeled on this dialog box in bold type. This indicates that these fields are required. The other fields can also be matched to return more accurate results in the location search. For more information regarding field requirements for the many address locator styles, see the chapter 'Preparing for geocoding' in this book.

Building an address locator for alternative searches

Creating address locators that will use alternative searching techniques is not very different from creating ones that do not. There are, however, some elements that must be included. By combining these tasks, an address locator can be created that includes alternative names and place name alias material.

Building an address locator for alternate names

1. In ArcCatalog, click the Address Locators folder at the top of the Catalog tree, then double-click Create New Address Locator.
2. Select an address locator style that includes "with AltName" in the description.
3. Click OK.

   The New Address Locator dialog box will appear.
4. In the Name text box, replace the default name with a desired name.
5. On the Primary table tab, click the Browse button next to the Reference Data text box and navigate to the shapefile or feature class you wish to use as your primary reference data, then click Add.
6. Click the Alternate Name table tab. ►

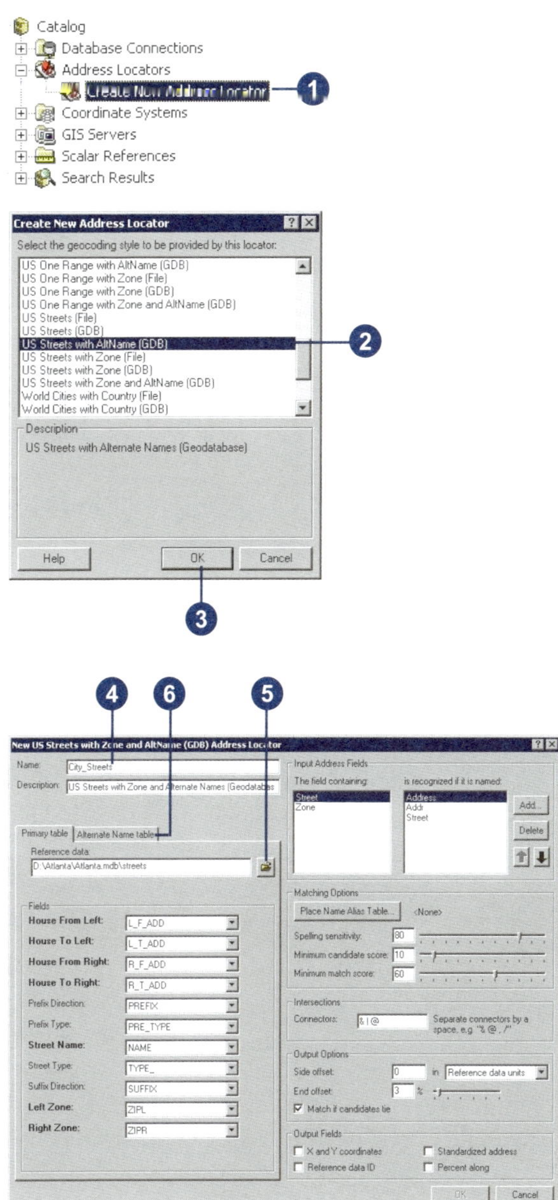

Tip

Alternate Name table field requirements

The required fields in the Alternate Name table will vary based on the address locator style used in developing the address locator. In addition to the style-specific requirements, each Alternate Name table must contain a column containing the alternative name as well as a column containing a Join ID value, referencing the entity back to the primary table.

For more information regarding reference data requirements for the variety of address locator styles, see the chapter 'Preparing for geocoding' in this book.

7. Click the Browse button next to the Reference data text box and navigate to the table you wish to use for your alternative name data, then click Add.

   This will close the navigation dialog box.

8. In the Fields section of the Alternate Name table tab, click the Join ID dropdown list and select the table field that will be used to join the table and the shapefile or feature class.

9. If needed, make any modifications to the address fields.

10. If desired, modify the address locator to other desired settings.

11. Click OK on the New Address Locator dialog box.

    Your new address locator will appear in the address locator folder in ArcCatalog and is ready to use for geocoding.

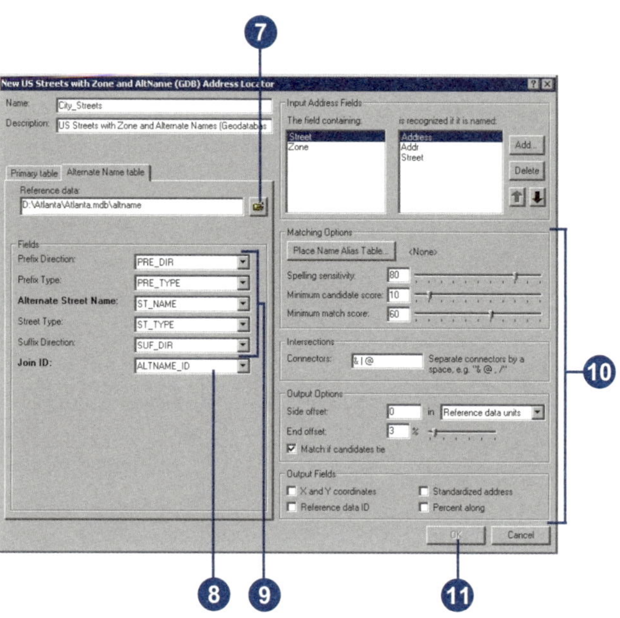

Building an address locator for place name aliases

1. In ArcCatalog, click the Address Locators folder at the top level in the Catalog tree, then double-click Create New Address Locator.
2. Select an address locator style.
3. Click OK.

   The New Address Locator dialog box will appear.
4. In the Name text box, replace the default name with a desired name.
5. On the Primary table tab, click the Browse button next to the Reference Data text box and navigate to the shapefile or feature class you wish to use as your primary reference data, then click Add.
6. Click the Place Name Alias Table button.

   This will open the Alias Table dialog box. ►

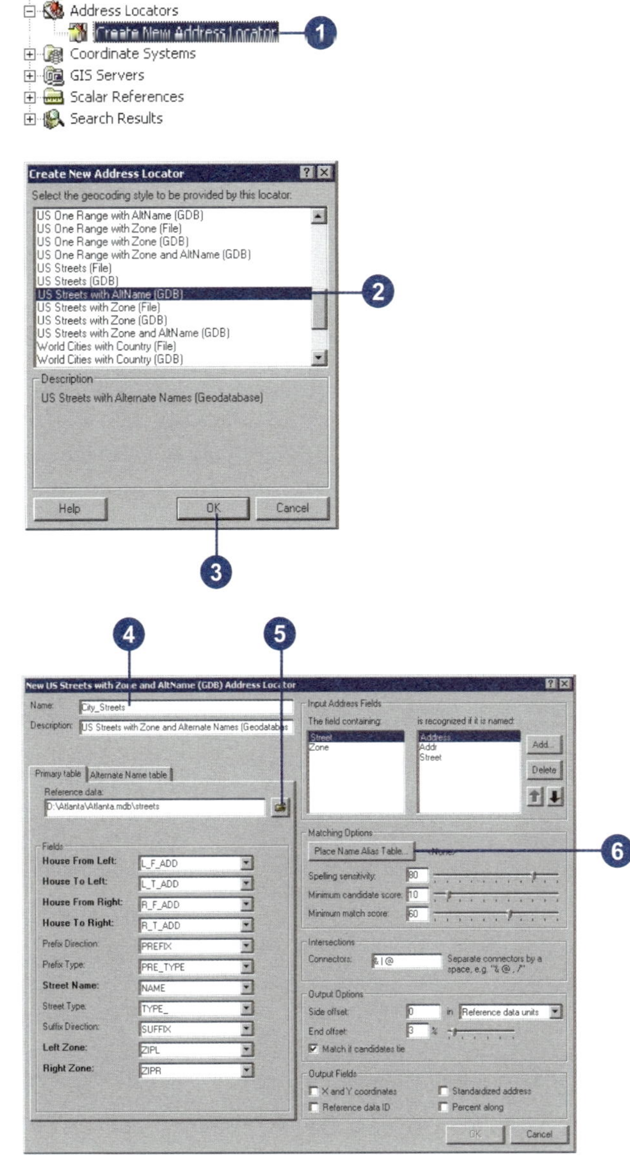

Tip

Place Name Alias Table requirements

The Alias Table has several required fields. The first of these is the Alias Field. This field contains the common name of the location. The other required and optional fields are based on the style of address locator you are creating.

For more information regarding field requirements for the variety of address locator styles, see the chapter 'Preparing for geocoding' in this book.

7. Click the Browse button next to the Alias Table dropdown menu.
8. Navigate to the table you wish to use for your alias name data, then click Add.

   The name of the alias table will appear in the Matching Options section of the New Address Locator dialog box.
9. Click the Alias Field drop-down menu and click the field that corresponds to the location or alias name.
10. If needed, make any modifications to the Address Fields, then click OK.

    This will close the Alias Table dialog box.
11. If desired, modify the address locator to other desired settings.
12. Click OK on the New Address Locator dialog box.

    Your new address locator will appear in the address locator folder in ArcCatalog and is ready to use for geocoding.

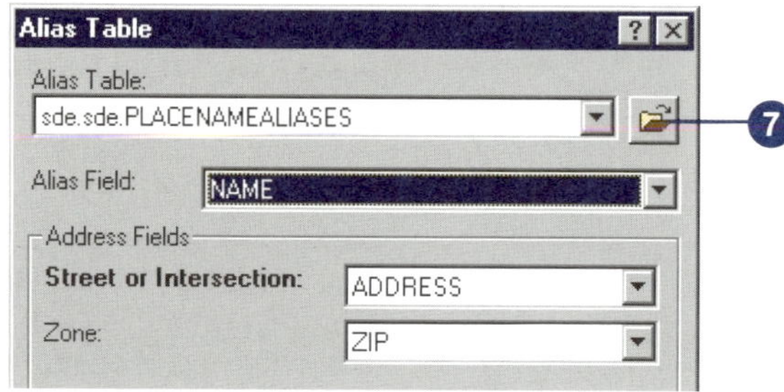

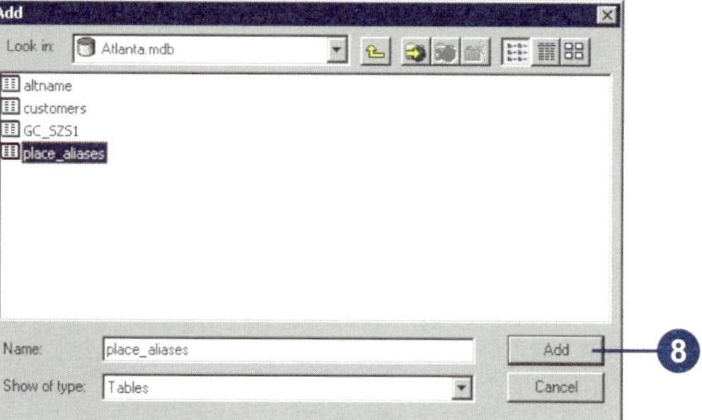

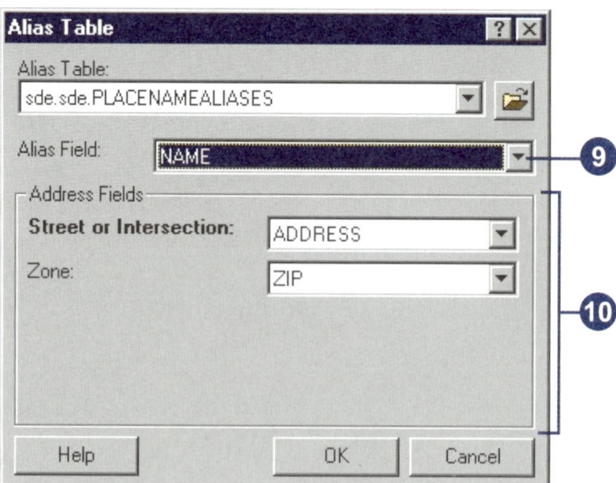

Searching for an intersection

In many geocoding applications, the address of a location is not as clear as the intersection of two streets. As discussed in previous chapters, you can search for a location based on the building number and street name. You can also search for a location based on the intersection of two line features.

Tip

Adjusting the match options for your address locator

If you find that your intersection is not matching, you can adjust the match options directly from the Find dialog box. Clicking the Options button on the Address tab will open a dialog box allowing you to make the needed adjustments.

1. In ArcMap, on the main menu, click Tools, then Geocoding, then Address Locator Manager.
2. Click Add.

   The Add Address Locator dialog box will open.
3. Navigate to the desired address locator, select it, and click Add.

   The address locator will be added to the Address Locator Manager.
4. Click Close to close the Address Locator Manager. ▶

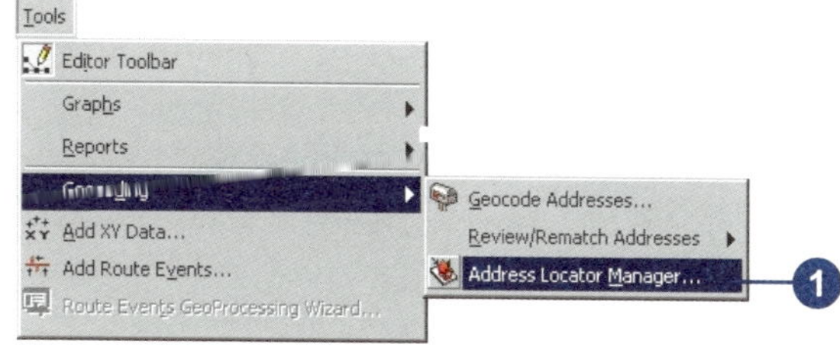

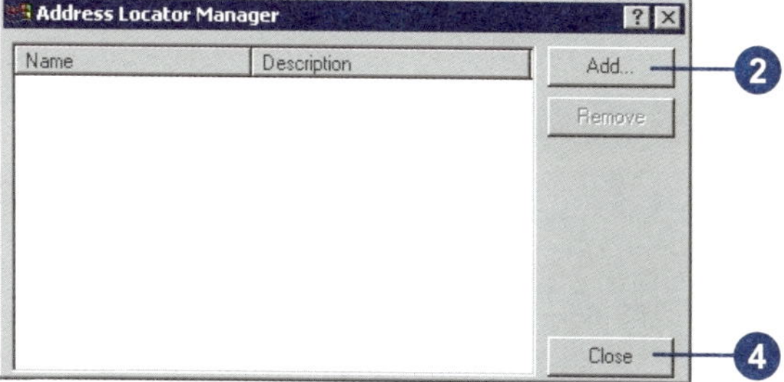

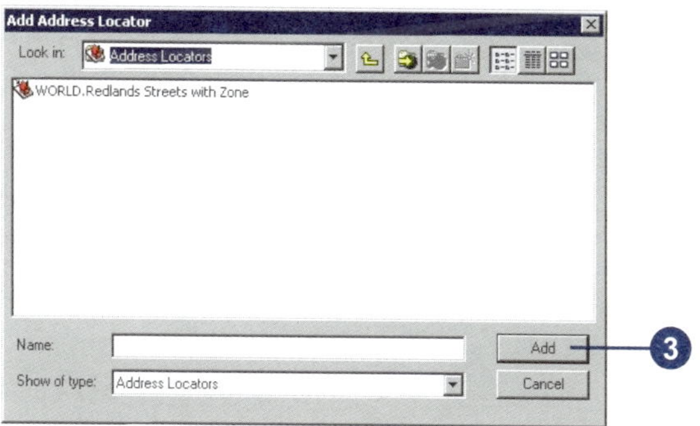

Tip

Improving your match score

Depending on the type of address locator style used, you may find that the intersection geocoding search is returning low match scores. If this is the case, try using additional street attributes, such as street type and direction, in your search parameters.

Tip

Batch geocoding using intersections

Just as you can search for a table of addresses, a table containing addresses and intersections or solely intersections can be searched in the same manner.

For more information on batch geocoding, see the chapter 'Locating addresses' in this book.

5. On the tools menu, click the Find icon.

   The Find dialog box will open.

6. On the Addresses tab, choose the appropriate address locator.
7. In the text box labeled Street or Intersection, enter the first street name, a space, a dividing symbol as set on the Address Locator Properties dialog box, another space, and finally the second street you wish to search.
8. Click Find.

   A list of candidates will appear in the lower portion of the Find dialog box.

9. Navigate through the list of candidates and take action on the appropriate match.

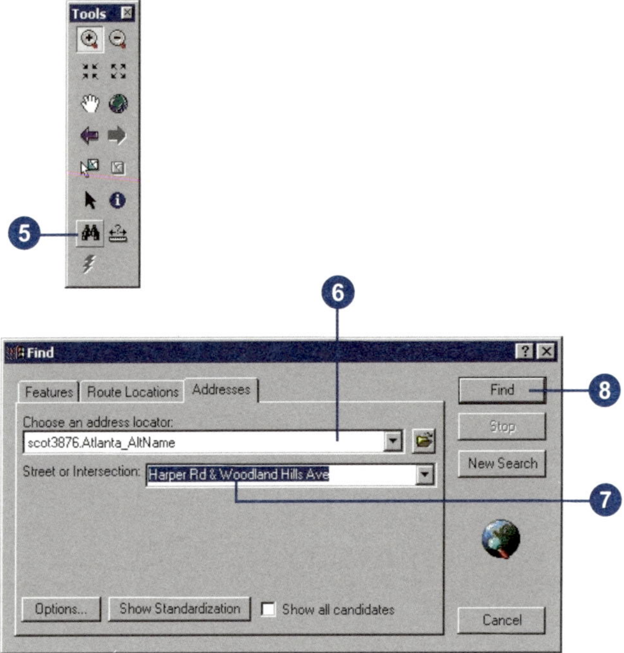

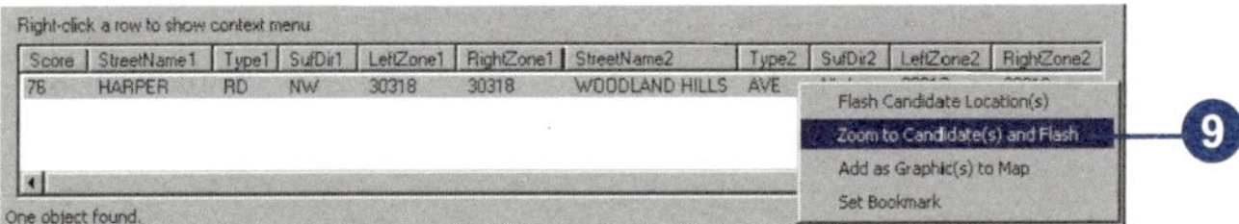

Keeping an address locator current 8

IN THIS CHAPTER

- **Updating your reference data**
- **Versioned data and the geocoding index**
- **Rebuilding geocoding indexes**

As with any data collection, editing and updating your reference data is a common task. The addition of a new subdivision to a city, a change in street names, the addition of new schools or hospitals, or simply a more accurate measurement of the existing data all constitute the need to update your reference data.

One important aspect when editing your reference data is the understanding of the geocoding index created at the time when the address locator was created. As you make modifications to your reference data, the related geocoding index must reflect those changes.

This chapter will review reference data and the changes to the reference data that you may make. Methods of maintaining the geocoding index will then be described.

Updating your reference data

A review of reference data

As discussed previously in this book, primary reference data consists of three distinct, yet related entities: the geometry, the attribute table, and the geocoding index. The geometry defines where in geographic space the items are located. This is usually presented in the form of a map. As the roads, parcels, or other features are presented, the arrangement is based on the geometry of the entities. The attributes associated with those entities are also an important part of the reference data. It is against many of the attributes of the reference data that the addresses are geocoded. The third element of the reference data is the geocoding index. As described previously, the geocoding index is created to facilitate and speed up the geocoding process. Certain elements in the attribute table are assigned specific index values. For example, alphanumeric values are assigned to each street name. The geocoding engine uses these index values to perform a quicker search rather than the traditional technique of searching through the entire attribute table.

Secondary reference data can also be applied to the address locator. This includes Alternate Name tables and Place Name Alias tables. These tables are described in the chapter 'Additional geocoding techniques' in this book. The production of indexes for the Alternate Street Name table is similar to the indexes created for the primary reference data attribute tables.

Common changes to reference data

Most organizations update or change the geometry and attribute table related to the reference data on a regular basis. This is a basic part of keeping the data current. New subdivisions or streets might be added. Alternate names might be used for streets, or a location name might have become commonplace. Additional regions or districts might be added to the database or previous errors may have been corrected. Each of these changes will alter the reference data associated with the address locator. For more information about editing shapefiles and feature classes, see the book *Editing in ArcMap*.

Updating your geocoding index

While it is desirable to modify, correct, or otherwise augment your map data, this can produce some complications in terms of geocoding. Bear in mind that the geocoding index was created specifically for the reference data that was present at the time the address locator was created. If your address locator uses feature classes and tables in a geodatabase as reference data, then the geocoding indexes are implemented as tables in the same geodatabase. When using shapefiles or coverages, the geocoding indexes are file-based and are stored within the file where the reference data is located.

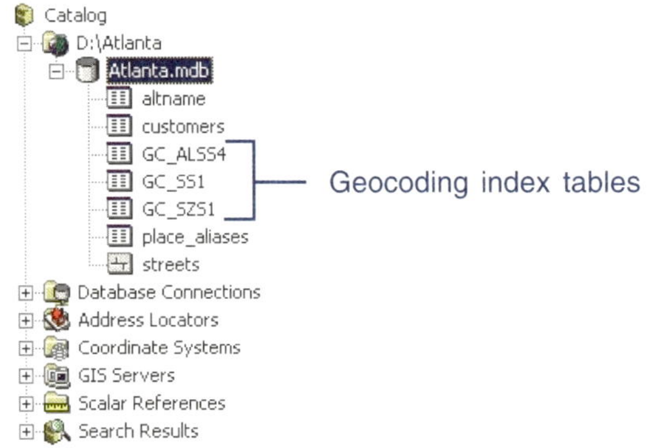

An updated relationship between the reference data and the geocoding indexes must be maintained in order to correctly geocode addresses. Several tools for maintaining geocoding indexes are described later in this chapter.

Versioned data and the geocoding index

When using versioned data within an ArcSDE geodatabase, the address locator is always created on the default version of the data. The geocoding index reflects the condition of the data in the default version. Maintaining the geocoding indexes within a versioned geodatabase can produce varying results based on the order in which you version and automate the geocoding indexes.

No automation of the geocoding index

Should you choose to create an address locator based on a versioned database but you do not automate the rebuilding of the geocoding index, the only version of geocoding indexes that exists corresponds to the default version, regardless of other versions of the data created.

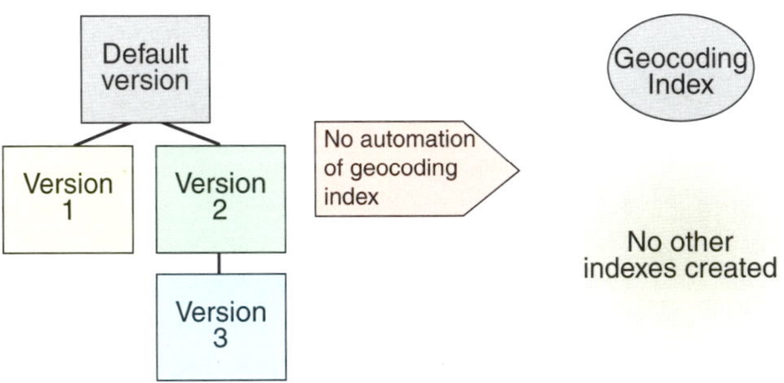

Without automating the rebuilding of your geocoding index, only one geocoding index is created that corresponds to the default version of the geodatabase.

Automating the geocoding index maintenance after the data is versioned

When you version your data, then automate the rebuilding of the geocoding index, an additional geocoding index is created for each data version. However, these indexes that are created are identical to the default version index regardless of any changes that you might have made to the different versions. As you make changes to the versioned data, those changes are maintained in the index created for that version.

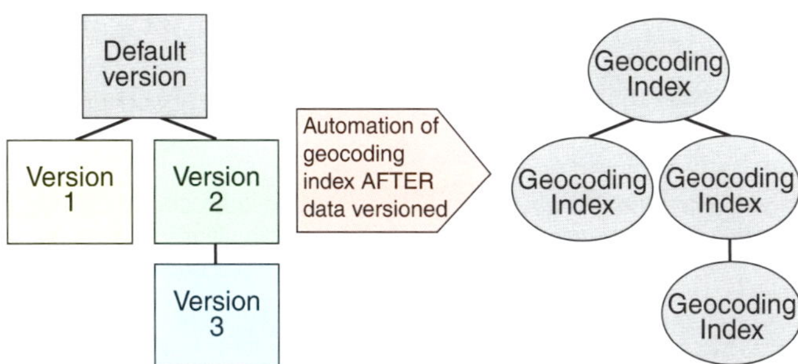

When you automate the maintenance of the geocoding index after the data is versioned, geocoding indexes are created; however, they do not reflect the versions for which they were created. Subsequent changes to the data are maintained in the geocoding index.

In order to geocode using a particular version, the changes need to be posted back to the default version. In so doing, the default version with its corresponding geocoding index will contain all of the modifications that were made to that particular version. Should you choose to post changes back to the default from more than one version, any discrepancies between the versions will need to be reconciled.

Automating the geocoding index maintenance before the data is versioned

Automating the rebuilding of the geocoding index before the data is versioned results in a geocoding index for each version that corresponds to the modifications that were made to each version.

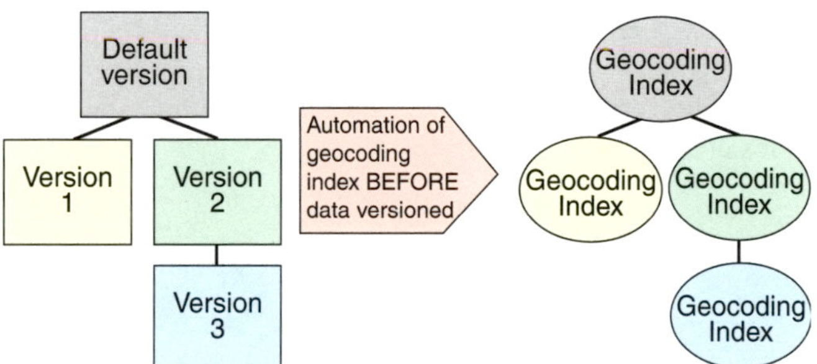

When you automate the rebuilding of the geocoding index before the data is versioned, as you version the data, a new geocoding index is created for each version. These indexes reflect the changes that you make to each version.

To geocode against a particular version, you can post the changes back to the default version or you can connect to that version and export the data to another geodatabase. This will produce another geodatabase containing all of the reference data based on that particular version.

Rebuilding geocoding indexes

When you make edits to your geocoding reference data (either by adding or deleting features or rows, or by editing the street names that they contain), you may need to update your geocoding indexes.

To facilitate the process of updating your geocoding indexes, the toolboxes provide several tools to perform this function. This can be set as an automatic procedure, or you can rebuild your index manually.

Manually updating the geocoding index

1. In ArcToolbox, open the geocoding toolbox and double-click the Rebuild Geocoding Index tool.

   This will open the Rebuild Geocoding Index dialog box.

2. Click the Browse button next to the Input Address Locator text box and navigate to the address locator that corresponds to the reference data you have changed.

3. Click OK on the Rebuild Geocoding Index dialog box.

   This will initiate the rebuild process.

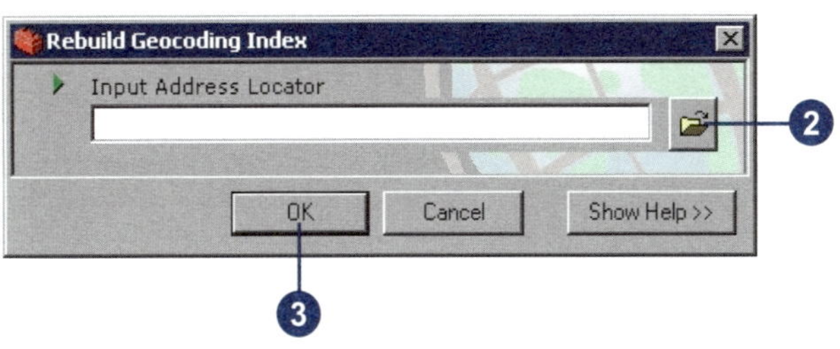

Tip

Concluding the automated process

To conclude the automated process, use the Unautomate Geocoding Indexes geoprocessing tool.The dialog box will appear the same as the Automate Geocoding Indexes dialog box.

Automatically updating the geocoding index

1. In ArcToolbox, open the geocoding toolbox and double-click the Automate Geocoding Indexes tool.

   This will open the Automate Geocoding Indexes dialog box.

2. Click the Browse button next to the Input Address Locator text box and navigate to the address locator that corresponds to the geocoding indexes you wish to automate.

3. Click OK on the Rebuild Geocoding Index dialog box.

   This will initiate the rebuild process each time modifications are made to the reference data.

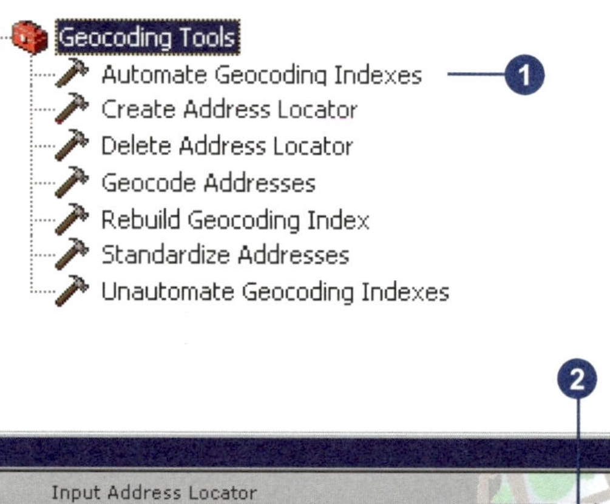

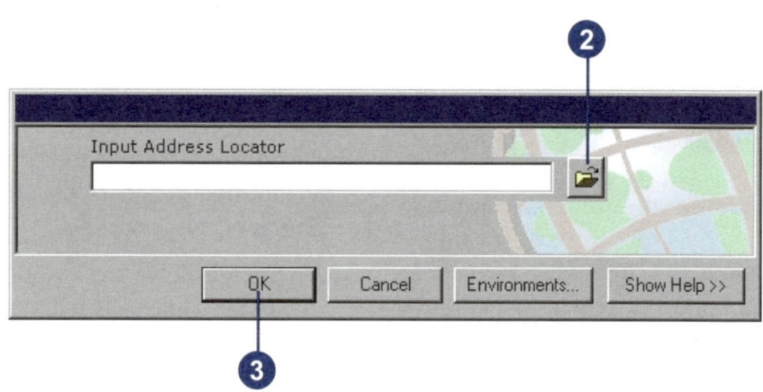

Distributing your address locator

IN THIS CHAPTER

- **Sharing elements of the geocoding framework**
- **Deciding to share an address locator**
- **Collecting address locator-related material**
- **Making your address locator public**
- **Accessing shared address locators**

As you create new address locators, you may want to share them with other users. This might include distributing the address locator within your organization or giving it to others who are collaborating with your research or study. Whatever the reason, there are a number of methods to distribute the address locator.

Sharing an address locator involves sharing the locator file as well as the address locator-related materials. This includes the reference data and geocoding indexes as well as any custom rule base files.

This chapter discusses the advantages and drawbacks of a variety of distribution techniques and explains some of the tasks involved in sharing geocoding-related files.

Sharing elements of the geocoding framework

As you may remember from the chapter 'Concepts of geocoding' in this book, the geocoding framework consists of several major components. These include the reference data and address style-specific guidelines as the two parts of the address locator, the geocoding rule base, and the geocoding engine. One of the first steps in preparing to share an address locator is determining what material needs to be accessible to the user, where this material is to be stored, and what elements of the geocoding framework the user already has. This section will review the geocoding framework and provide further details on the format of each of the geocoding framework elements.

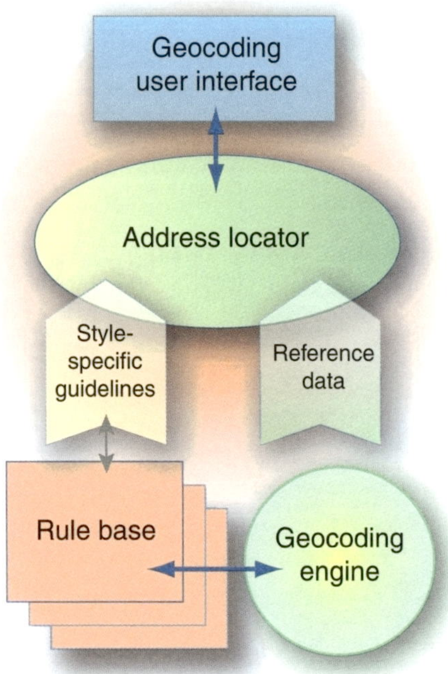

When preparing to distribute an address locator, it is important to remember each of the elements of the geocoding framework.

The address locator

The address locator consists of two major entities. These are the reference data specific to the region where addresses will be geocoded and the style-specific guidelines.

Reference data

As discussed in previous chapters, the reference data consists of several different parts. The first part of the reference data to be considered is the primary reference data. Along with the primary reference data is the geocoding index created at the time the address locator was built. The primary reference data consists of a basic shapefile or a feature class within a geodatabase. The geocoding index is simply a table. When a feature class within a geodatabase is used as the primary reference data, the geocoding index is stored within the same geodatabase. When a shapefile is used, the geocoding index is stored within the same folder as the shapefile being used.

Apart from the primary reference data and related geocoding index, there are potentially several additional pieces of ancillary or secondary reference material. These include the material related to the Alternate Name table and the Place Name Alias table. Each of these file types are also, simply, tables. Using these tables, geocoding index tables are also created.

When distributing an address locator, it is important to understand that all of these files are crucial elements in a address locator function. Without access and reference to these files, the address locator will not function.

Address style-specific guidelines

The guidelines that direct the geocoding engine during the geocoding process are contained in an individual file referred to as the address locator. This file stores all of the style-specific guidelines as well as the settings you adjust on the Address Locator Properties dialog box. A new file is created for each individual address locator and is distinguishable by the file extension .loc. When distributing an address locator, this file is the backbone of the address locator.

The geocoding rule base

The rule base is a collection of files that directs the geocoding engine in how to standardize address data and match it to the related location in the reference data. Each address style uses a specific set of rule base files. The functions of the rule base are twofold: the first is the address data standardization process; the second is the matching of the address to the reference data.

Standardization is a process of address parsing that prepares the address to be matched against the reference data. It converts the commonly abbreviated address categories to standardized values, for example, "Blvd" to "Boulevard", and breaks down the address into distinct address categories referred to as match keys.

The standardization files include the standardization commands, stored in a standardization file; a match key dictionary for the event address, stored in a match key dictionary file; a classification table for providing standard address abbreviations, stored in a classification file; and patterns and actions for standardizing an address, stored in an action–pattern sequence file, referred to as a pattern file. The standardization command file specifies input record size, debug mode, output filename, and the process name. The match key dictionary defines the data type, field length, and missing value code for each match key field. The classification table, in turn, interprets various abbreviated address elements and provides a standard abbreviation for each. The pattern file recognizes certain patterns in the address format and dictates to which address field each element of the address format should be assigned.

Match rules are used to define the methods of address-to-reference comparisons and have two major functions. First, this file, termed a match file, defines what address elements the fields in the reference data attribute table represent. The match rule file also specifies probabilities for score comparison between the address elements in the reference data and the address elements in the address being searched. These probabilities are known as the m and u probabilities.

Specific detail regarding each of these files is provided in the *Geocoding Rule Base Developer Guide*. In distributing an address locator, you must bear in mind that these files need to be accessed, either indirectly off a server-side machine or directly on the client machine.

The geocoding engine

The geocoding engine consists of several files within the ArcGIS program files. As mentioned previously in this book, this file reads the directions dictated by the address locator file and processes the address being searched against the functions in the rule base, resulting in a location match for the address.

When distributing an address locator, a specific geocoding engine must be indicated. This can be a geocoding engine on a server or the engine included on the client machine.

Deciding to share an address locator

Before taking action in collecting the geocoding-related files and publishing them, it is important to ask several questions about the purpose and audience of your distributed address locator. This will focus your distribution efforts and simplify the process of distribution and access.

What do others want to do?

One key concept in distributing your address locator is in knowing what the user wants to do. This will range from creating a new address locator based on a specific, customized address locator style to geocoding a table of addresses against a customized address locator and modified rule base files to looking for a single address with a predefined address locator. An understanding of what the users want to do will help you in determining what parts of the address locator you need to make available to them.

With whom are you sharing?

Another important consideration is knowing with whom you will be sharing your address locator. The main importance of this question is in knowing what parts of the geocoding framework the user already has. For example, if the user already has ArcGIS Desktop, all of the standard rule base files are already available. You will not need to distribute these, unless the potential user does not have the rule base files that an address locator might use, such as customized rule base files. Also, the potential user might already have a copy of the reference data. Alternatively, the user might be connected to your same local network and will be able to access all of the reference data, as well as the address locator-related files, directly from your computer.

How do you want to share your address locator?

There is a wide range of distribution methods to share an address locator. The process, as well as the advantages and disadvantages for each, is described below.

Local area network

Distributing your address locator over a local area network, or LAN, allows other people connected to your local network to access files on your computer. By granting permission for others to access these files, they are able to network into your computer and, based on the degree of permissions granted, read or write over these files.

Sharing your geocoding service over a local area network is one of the simplest methods of distribution. In essence, you could simply share the folders on your computer where any files related to geocoding are stored. This would allow others to access these files at any time. This method of mass file sharing is not advisable. It makes your computer vulnerable to mischievous actions and could exhaust your computational resources.

Instead of sharing all geocoding files, you could copy all of the files related to one address locator into one folder and share that individual folder. The process of collecting the address locator-related files is described in the section 'Collecting address locator-related files' in this chapter. Granting permission to your files for others to access varies slightly based on your network configurations and operating system. Consult with your system administrator about this process.

While the process of sharing is easy, there are several drawbacks to distributing over a local network. One disadvantage is that those outside or disconnected from the local network are not able to access the needed files. When traveling or working away from the office, the user might not be able to access the needed files and the address locator will fail. Also, when others access your computer, this requires the attention of some of your computer's resources. If a significant number of people are going to be accessing or processing from your computer, this might exhaust the computational resources needed to perform basic tasks.

Packaging onto a compact disc or compressed file

In the section of this chapter 'Collecting address locator-related files', steps are described to collect all of the needed material to distribute an address locator. In essence, once these steps have been performed, the file populated with these files could be burned to a compact disk or put into a compressed file format.

Obviously, this distribution technique is easy. Further, the address locator can be included in other packages of geographic data.

One disadvantage of this type of distribution is that you are limited in distribution to the number of compact discs you produce with the needed information or the method that you use to distribute the compressed file. In order to share the service with a large number of users, you will need to give each user access to the packaged material.

The material that you package and distribute might also become outdated. In order to provide the users with a current version of the data or address locator, you will need to reproduce and redistribute the material each time it is updated.

ArcSDE

The creation and access to ArcSDE address locators is similar to the interface used in client-side address locators. For this reason, these services are relatively easy to create and distribute. The reference data is stored on the same server, while the processing takes place on the ArcSDE server machine.

This type of distribution does have some requirements that your users might not have access to. First, to create or access the ArcSDE server, you must have ArcInfo or ArcEditor. ArcView users have read-only access to the server. Also, any potential users need to have permission granted to access the data and service in the ArcSDE server.

ArcGIS Server

Another method of distributing an address locator is through an ArcGIS Server. ArcGIS Server allows you to publish address locators as "services". Depending on your application, this can have a great advantage over other methods of distribution. The service can be distributed to a wide range of users and, with a standardized communication format, the service can be accessed for a wide range of applications. These services can also be used directly in ArcGIS applications.

As with ArcSDE, one drawback of distributing with ArcGIS Server is that some administrative processes must be done to establish and manage these services.

For more information about ArcGIS Server and its applications, see the *ArcGIS Server Administration and Development Guide.*

Collecting address locator-related material

Once you have determined how you wish to distribute your address locator, the material preparation and management should be performed in ArcCatalog. Depending on the source and format of the material, distribution techniques will vary. Each of these variations is illustrated in the following table and described below.

Geocoding Elements	ArcView 3	ArcGIS (file-based)	ArcGIS (GDB-based)	ArcSDE
Address Locator	Automatically copied when reference data is copied	Using ArcCatalog, copy into public folder.	Using ArcCatalog, copy into public folder.	Publicly available
Reference Data	Using ArcCatalog, copy reference data into public folder. The locator and related geocoding indexes will copy automatically.	Using ArcCatalog, copy reference data into public folder. The related geocoding indexes will copy automatically.	All of the reference data and the geocoding indexes must be in the same geodatabase. To distribute the geodatabase you can: · Copy the entire GDB or · Export the reference data and geocoding indexes to another geodatabase.	Grant SELECT privileges to consumer.
Geocoding indexes	Automatically copied when reference data is copied	Automatically copied when reference data is copied		Grant SELECT privileges to consumer.
Geocoding Rules	located in: ArcGIS/Geocode	located in: ArcGIS/Geocode	located in: ArcGIS/Geocode	Publicly available

ArcView 3

If you have address locators that were created using ArcView 3, distribution of the address locator and the associated file-based data source is rather straightforward. You simply need to copy and paste the reference data into a public folder. In doing so, all of the associated files, including the index and the address locator, will be copied as well.

Regardless of the application used to create the address locator, if you have created custom geocoding rules that you also wish to distribute, these too can be copied into the public folder. These files do not appear in ArcCatalog and will need to be distributed using a more general file management software, such as Windows® Explorer. In order to determine which files are needed, you will need to open the address locator file in a text-editing software, such as Microsoft® Notepad.

In ArcView 3 the file extension of the address locator file will vary depending on the data type of reference data used. The following table will make finding the correct file easier.

Reference data	Locator	Geocoding index
Shapefiles		
all	.mxs	.ixs
Coverages		
Point	.mxx	.ixx
Arc	.mxa	.ixa
Polygon	.mxp	.ixp
Region	.mxr	.ixr
Route	.mxd	.ixd
CAD		
Line	.mxo	.ixo

When geocoding in ArcView 3, the file extension of the address locator file and the geocoding index varies based on the type of reference data used.

Open the address locator file that was automatically copied into the public folder. In that file, you will find a long list of key-value properties of the address locator. Using the Find function in the text software, locate the 'MatchRules' property. The value associated with this property is the name of the .mat rule base file. Make note of that name and search for the 'StanRules' property. The value associated with this property is the name of several files that you will need.

The geocoding rule base files are, by default, stored in the ArcGIS/Geocode folder where you stored the ESRI software components. Once this file is found, you will need to copy the .mat file into the public folder. The name that was associated with the .stn file has several files that share that same name. All of these files need to also be copied into the public folder.

For example, if the address locator file contained the following two key-value properties,

```
MatchRules = us_addr1.mat
StanRules = us_addr.stn
```

then you would need to copy the following files into the public folder.

- us_addr1.mat
- us_addr.stn
- us_addr.pat
- us_addr.cls
- us_addr.dct

Before using this address locator, the client who will be using these rules should copy them into his/her own ArcGIS/Geocode folder.

Whenever distributing an address locator, the final step is to open the address locator properties dialog box and change the path names to the reference material that has been copied into the public folder. This process is described later in this chapter.

ArcGIS file-based reference data sources

With some variation, the process of distributing an address locator created in ArcGIS, using file-based reference data, is similar to the process described for ArcView 3 address locator distribution.

The address locator file can be stored anywhere on the file system. However, by default, when creating an address locator, it is stored within the user profile directory. When managing these files within ArcCatalog, the address locators that are stored within the user profile directory will appear in the Address Locators folder in the Catalog tree. These address locator files can be moved to any other file or directory through copy and paste commands done in ArcCatalog.

In addition to moving the address locator file to the public folder, you will also need to move the reference data. When you copy and paste file-based reference data within ArcCatalog, the geocoding index that was created for that data will automatically be copied.

If you have created custom geocoding rules that you also wish to distribute, these should also be copied into the public folder. Again, these files do not appear in ArcCatalog and will need to be distributed using a more general file management software.

To determine which rule base files the address locator uses, open the address locator file that was copied into the public folder with a text-editing software. This file will have a .loc file extension. Just as in the ArcView 3 address locator file, you will find a long but slightly different list of key-value properties. The properties that define the .mat file and the .stn file are 'FileMAT' and 'FileSTN', respectively. Using the Find function in the text software, you can locate these properties and make note of their names.

With these names, you can go to the ArcGIS/Geocode folder and copy the needed files. Again, the name that was associated with the .stn file has several files that share that same name. All of these files need to be copied into the public folder.

Any custom rule base files should be copied into the user's ArcGIS/Geocode folder.

As with the ArcView address locators, the final step is to open the Address Locator Properties Dialog box within ArcCatalog and change the pathnames to the reference material that has been copied into the public folder. This process is described later in this chapter.

ArcGIS geodatabase reference data sources

As when using file-based reference data, the process of moving an address locator file from one location to another is identical when using geodatabase-stored reference data.

When distributing the reference data, the reference data is stored as a feature class and a series of tables within the geodatabase. These additional tables include the reference data index tables. Depending on their usage, these tables might also include an Alternate Name table with its index or a Place Name Alias Table. The Place Name Alias Table does not use an additional index table.

To distribute this material, you may wish to copy and paste the entire geodatabase, or you may wish to create a new geodatabase and export the reference data and its related index tables, or an entire feature dataset from one geodatabase, into the newly created geodatabase.

You can find the names of the index tables created within the address locator file. Open the address locator file that was copied into the public folder with a text-editing software. This file will have a .loc file extension. Search for the RD.Val.Table property. This property key is often followed by a number, for example,

```
RD.Val.Table1 = Streets
```

For each table property, you will also find a corresponding index table property. These are the RD.Value.IdxTable properties. This property value is the name of the index table that was created within the geodatabase for that reference data. Information regarding the creation of geodatabases and exporting material from them is provided in *Building a Geodatabase*.

The process of distributing customized geocoding rule base files for geodatabase address locators is identical to the process described previously for file-based address locator rules.

ArcSDE

If you have created an address locator within an ArcSDE geodatabase, all of the needed files are already within the database. Locators and geocoding rules are accessible to all ArcSDE users; however, you will need to modify the privileges for the reference data and the associated index tables to allow others to use the address locator against that data.

Making your address locator public

There are several common steps needed to make an address locator accessible to others. These are presented in this section.

Tip

Using relative pathnames

If the file relationship between your reference data and the address locator file remains constant, you may wish to simply use relative pathnames. This option is available by checking the checkbox related to the primary reference data on the Address Locator Properties dialog box.

Changing the file pathnames in your address locator

1. In ArcCatalog, open the Address Locator Properties dialog box.
2. Click the Browse button next to the Reference data text box.
3. In the Navigate dialog box, navigate to the primary reference data that will be used for the address locator.

   This reference data must have the same availability as the address locator and must be stored in conjunction with its related index file.
4. If an Alternate Name table is used, repeat this process for the related reference data.
5. If a Place Name Alias table is used, repeat this process for the related reference data.

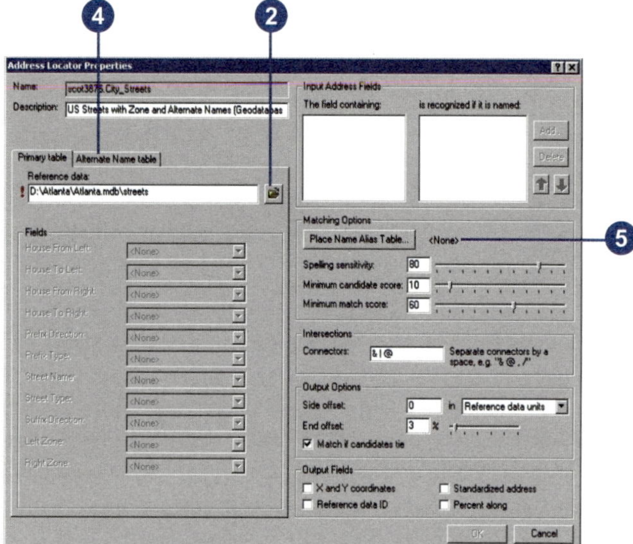

Sharing your address locator on a local area network

1. In ArcCatalog, right-click the folder that contains the address locator and related elements you wish to make public.
2. Click Properties.

   The Properties dialog box for the folder will open.
3. Click the Sharing tab.
4. Click the button indicated to share this folder.
5. Modify the share name or comments text box as desired.
6. Adjust the user limit as desired.
7. To specify particular users within the network who can access the files, click Permissions; otherwise skip to step 14.
8. On the Permissions dialog box, highlight and remove users or groups that you do not wish to have access to the file.
9. Click Add to add individuals or groups to the permissions list.

   This will open the Select Users or Groups dialog box. ▸

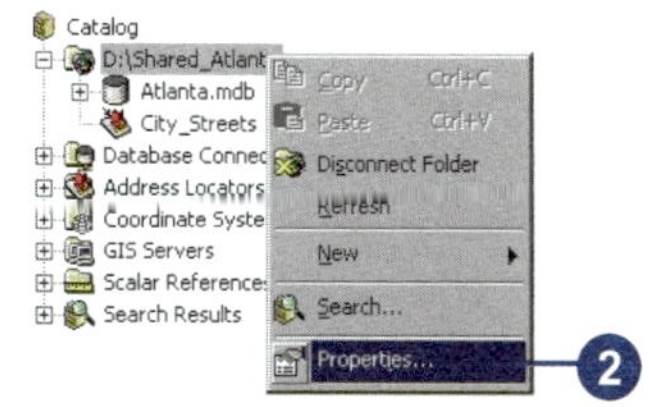

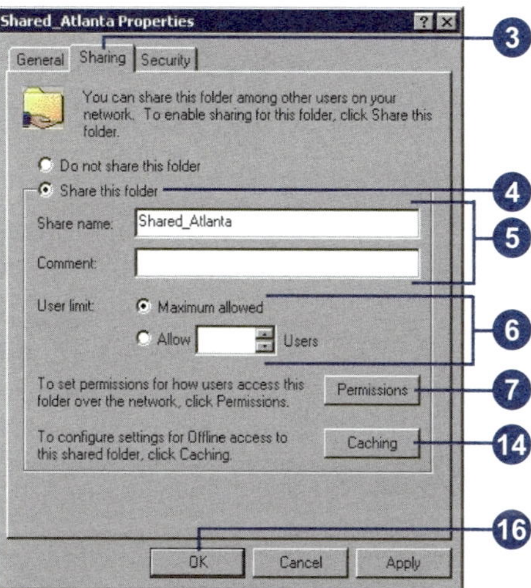

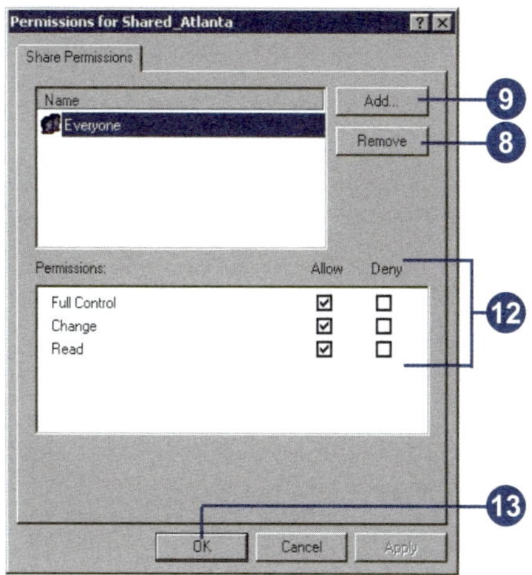

10. Select the user or group from the list provided in the upper portion of the dialog box and click Add.

    The name or names you add will appear in the lower portion of the dialog box.

11. Click OK to close the Select Users or Groups dialog box.

12. Adjust the permissions to allow or deny full control, change, or read options.

13. Click OK again to close the Permissions dialog box.

14. If you wish to modify the file cache process on the client-side machine, click the Caching button; otherwise skip to step 16.

    The Caching Settings dialog box will open.

15. Choose the setting that best accommodates your needs, then click OK to close the Caching Settings dialog box.

16. Click OK to close the Properties dialog box.

    The file will now be accessible to others within the local area network.

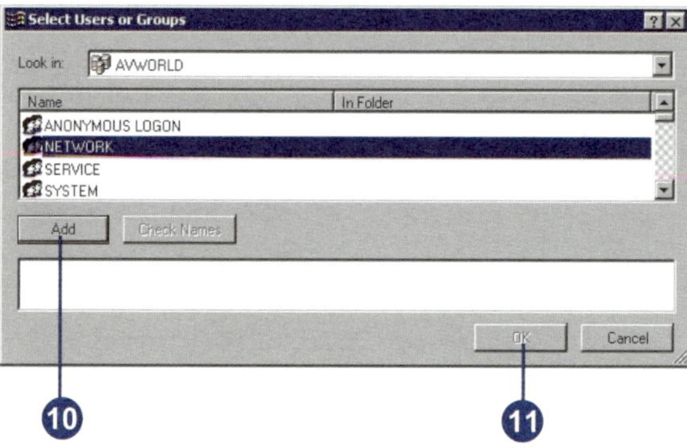

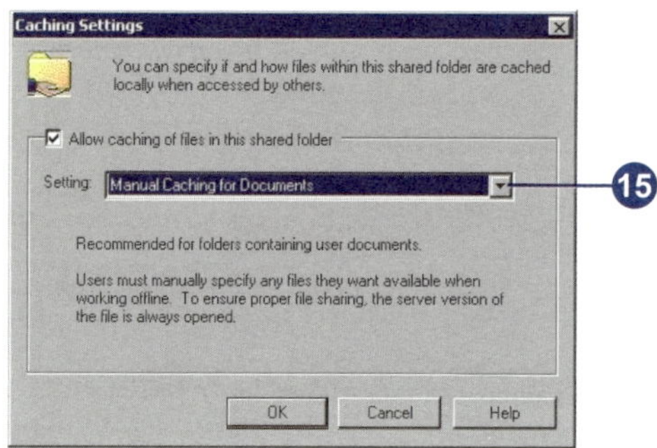

Tip

Compressed folders and compact discs

The actual process involved in creating a compressed folder or burning material onto a compact disc is specific to the software being used. This task simply describes the collection of material to include in the packaged format.

Distributing your address locator via a compact disc or compressed file

1. Collect all of the address locator-related files into a single folder or location.
2. With the address locator and all reference material in the same folder, correct the geocoding file pathnames to the reference data according to the task 'Changing the file pathnames in your address locator' in this chapter.
3. Open the software being used to create a compressed file or transfer material to a compact disc and follow the directions accordingly.

Tip

Building an ArcSDE address locator

With administration privileges to ArcSDE, building an ArcSDE geodatabase uses all of the same dialog boxes as when building a client-side address locator.

Sharing an ArcSDE address locator

1. Click the ArcSDE database connection that contains the address locator you want to make public.
2. Right-click the feature dataset that contains the feature class, the stand-alone feature class, or the table that the address locator uses as reference data and click Privileges.
3. Type the name of the user to whom you want to make the address locator available.
4. Check SELECT.
5. Click Apply.
6. Repeat steps 3 through 5 to make your address locator available to additional users.
7. Click OK.
8. Repeat steps 2 through 7 for the Alternate Street Name tables and Place Name Alias tables if your address locator uses them.

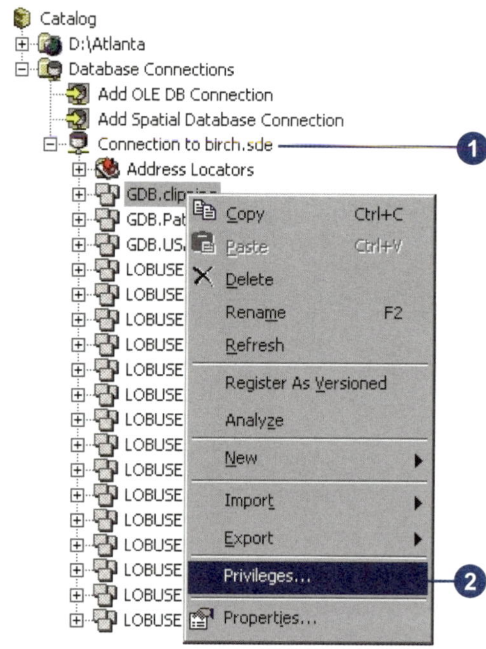

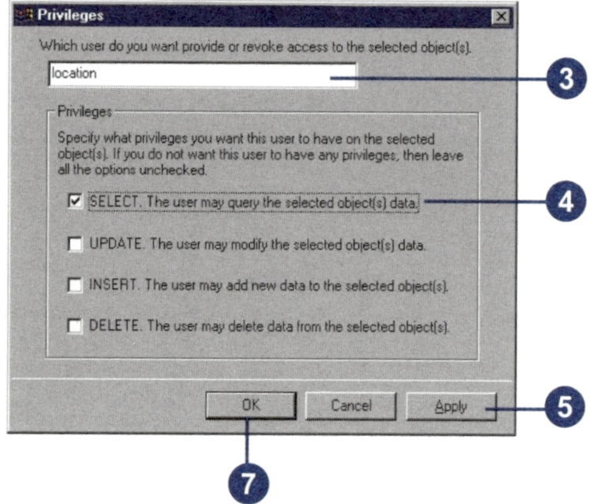

Accessing shared address locators

The process of accessing shared address locators is similar to the process used with those stored on your computer. ArcCatalog provides specific folders that are used to store connections to shared address locators. Each process is outlined in this section.

GIS Servers

Both ArcGIS and ArcIMS Servers can easily be added through the ArcCatalog tree. A simple interface allows you to insert the pathname or Internet address for the service you wish to add. These can then be used throughout ArcGIS.

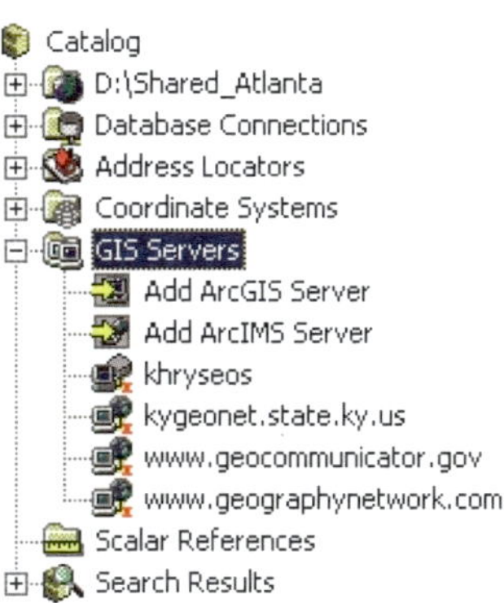

For more information on connecting to ArcGIS Server, see *Using ArcCatalog*.

Local area network

The file saving process from a local area network varies slightly according to the caching settings adjusted for the server-side computer file that stores the geocoding-related files. Although the file saving might vary, any publicly shared address locator can be reached through ArcCatalog. By clicking the Connect to Folder button, you navigate to all available folders, both on your own computer as well as those shared on other computers.

For more information about connecting to files, see the book *Using ArcCatalog*.

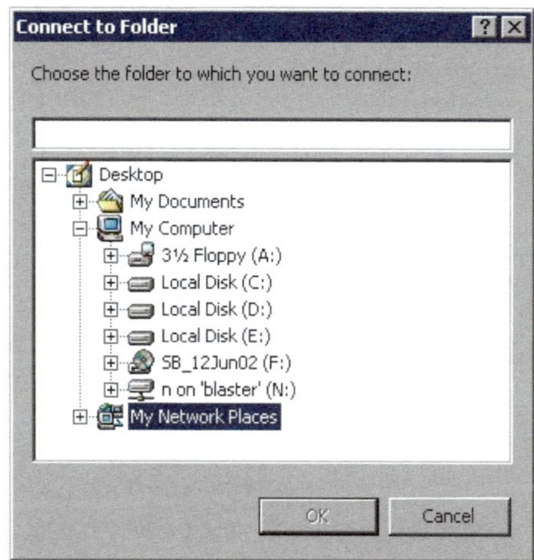

Compact disc or compressed folder

Material that has been stored on a compact disc is accessed as if it were already on your computer. In ArcCatalog, you simply need to navigate to the compact disc drive and use the address locator. You might find that the processing works more quickly, however, if you copy the files onto your hard drive. This would also allow you to access the address locator without the need of the compact disc.

When using a compressed file, you simply need to decompress the file and store the resulting folder on your computer. The address locator and reference material will then be available for use.

ArcSDE

If ArcInfo or ArcEditor is installed on your computer and you have access to an ArcSDE geodatabase, you can do server-side geocoding. Instead of creating address locators on your computer, you create and store them on the server. Server-side address locators are easily shared by many users in an organization. When you use them, the server will do the work of matching addresses to geographic locations.

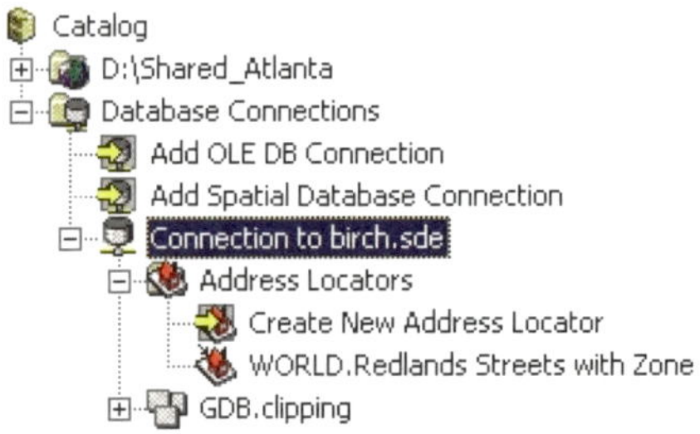

Once you have created a database connection that accesses the appropriate ArcSDE server, you can copy feature classes or tables from a personal geodatabase to the ArcSDE geodatabase.

If you choose to do so, create your address locators in the Address Locators folder within the database connection and use the data located in the same database. The resulting feature classes created by the geocoding process can also be stored within the ArcSDE geodatabase. Aside from these differences, the geocoding process will be the same as when you use client-side data and address locators.

Modifying your address locator 10

IN THIS CHAPTER

- **Exploring the address locator file**
- **Why change your address locator file?**
- **Changing default settings**
- **Table schema and the locator file**
- **Accommodating an alternative reference data schema**
- **Modifying the query**
- **Accommodating changes in the rule base files**
- **Special cases**

When creating an address locator, you begin by specifying an address locator style. Template files exist for each of the style types and have a file extension of .lot. When you create an address locator, the style-specific template is populated with the information that pertains directly to the address locator you are creating. These files have the file extension .loc.

This address locator file contains a wide range of settings used to direct the geocoding engine in the paths and processes necessary to translate the address information into a location. Many of these settings can be changed in order to direct the geocoding engine into using alternative methods or tools in translating the address information and interpreting a location. This type of altering is one of the more advanced methods of customizing your address locator.

This chapter begins by exploring a standard address locator file that would be used for creating an address locator based on reference data stored within a geodatabase. Reasons and applications for altering the address locator file are discussed, with several common tasks described. This chapter concludes by discussing some of the specific address locator issues that arise from the use of shapefiles as reference data or creating address locators within an ArcSDE geodatabase.

Exploring the address locator file

The address locator file has a wide range of sections, properties, and values. In this section you will be introduced to each of the sections and the function of these individual properties. This section should act as a reference for you in exploring the address locator properties. The section entitled 'Why change your address locator file?' will discuss the implications of making alterations to the template file.

Finding the address locator file

There are, in essence, two types of address locator files. These include template files for an address locator and the actual address locator files, with all of the reference data and other information included. By default, the template files are stored in the arcgis folder created on your hard drive when you installed ArcGIS Desktop. They are found in the subfolder 'Locators'. There you will find a collection of all of the template files that appear on the Create New Address Locator dialog box when creating new address locators. It is also here that you can store any address locator file templates you might modify or create.

Although the file extension for this file is .lot or .loc, you can open this file in any basic text editing software. As you proceed through this chapter, it is recommended that you open one of your address locators to see the different sections and properties described in the following material. This will also help you understand the process of customizing your address locator file. It is also advisable that, before you begin altering any address locator files, you make a copy of the original file in the event that your modifications cause undesired effects.

Some address locator file basics

Address locator properties are key–value pairs. Both keys (property names) and values are organized as strings. The maximum length of the property name or key is 32 characters and the maximum length of the property value is 255. The property keys are case insensitive and should not contain spaces. Both property keys and property values are separated by an equal sign:

```
MyPropertyKey = My Property Value
```

Spaces around the property value are stripped, so

```
MyPropertyKey=My Property Value
```

and

```
MyPropertyKey =   My Property Value
```

refer to the same property value, My Property Value. If it is desired to preserve leading and/or trailing spaces in the property value, double quotes should be used:

```
MyProperty =   "  My Property Value With Leading
and Trailing Spaces   "
```

Arrays of properties are represented by properties with the same name:

```
Fields = Address
Fields = City
Fields = State
Fields = Zip
```

Arrays of properties can be “spread out” among other properties; only the order between them matters. For instance, having

```
Fields = Address
Fields = City
AnotherPropertyKey = Another Value
Fields = Zip
```

or

```
Fields = Address
Fields = City
Fields = Zip
AnotherPropertyKey = Another Value
```

produces the same results.

Why change your address locator file?

Is changing the address locator what you need?

Before you begin the process of modifying or creating an address locator file, there are several considerations to make.

One of the first things to consider is what you are hoping to accomplish. What is not working correctly? What would you like to make work better? Would a modification make your work flow smoother? Do you find yourself needing to repeat the same modifications over and over? Do you have a special case scenario that you need to accommodate, such as a third party geocoding engine? If you have a clear goal in mind, you will be able to direct your efforts to the address locator file keys that pertain to your needed change.

It is possible that the modification that you want to make can be done in the Address Locator Properties dialog box. If this is the case, you may find that making modifications there will be more straightforward to understand than attempting to make the same change in the file.

On the other hand, if, when creating new address locators, you are needing to repeat the same modifications in the Address Locator Properties dialog box, then making the modifications to the address locator file template would be in order.

In considering what you hope to accomplish, you may find that the change that you need applies to other components of the geocoding framework and not the address locator file. This may include modifying the address of reference data or modifying the geocoding rule base. Often when modifications are made to the rule base, you must also alter the address locator file.

At times, the address locator file templates cannot accommodate the reference or address data format. Beyond the modification of an address locator template file, you can create additional address locator template files by writing one from scratch or by altering an existing file and saving it under a different name.

Changing default settings

You may find that when creating address locators, you are needing to make the same modifications on the Address Locator Properties dialog box. For example, your address data might have an additional symbol used to distinguish an intersection connector. Instead of making this modification with each address locator, you can make a general modification on the address locator template file. Once this is done, each time an address locator file is created using that template, the change will already be applied.

Accommodating alternative reference table schema

In ArcGIS there is a wide range of address locator templates available. These templates are designed to accommodate some predefined reference data schema including additional tables such as Place Name Alias tables and Alternate Name tables. You may find, however, that none of these template files accommodate the table schema for your reference data. For example, you may have your data organized in a specific address data model schema. You might also have some special cases where attributes are distributed across several tables, or several queries need to be performed to accommodate the material on a number of attribute tables. In these cases, you will need to modify the reference data schema laid out in the locator file.

Modifying the query

Along with modifying the table schema, you may need to modify the query process. One common example of a varied query process occurs when using an Alternate Name table. One query is performed to search for an address based on the names in your primary reference data. Another query is performed using the names found on the Alternate Name table. The final match scores of the candidates retrieved by both of these queries are the result.

There are several other applications in which you may need to modify the query process. For example, you may need to query an address against two feature classes, or you may wish to search for either a ZIP Code or for a city. This final example will be described in more detail later in this chapter.

Accommodating modifications to the rule base

As a more advanced method of customizing the geocoding process, at times you may need to make changes to the geocoding rule base files. Based on the modification that you made to the rule base files, some modification may also need to be done to the address locator file.

Changing default settings

When you create an address locator, there are a number of properties stored within the template file that dictate the default settings for many of the address locator properties. Depending on your reference or address data, you may need to make adjustments to these settings whenever creating an address locator. For example, your address data might use a symbol such as a plus sign "+" or a back slash "/" to separate intersections of streets. These symbols are not within the default settings for most address locators. If, when creating address locators, you find that you are needing to constantly make such adjustments, you may wish to make these changes within the template file. In this case you can create a template file that already recognizes the adjustments and will appear in the user interface.

To review, the address locator template files differ from the files created for individual address locators. The template files used in creating client-side address locators use the file extension .lot and are stored in the ArcGIS/Locators folder. In addition to these files, you will find files with the extension .avs. These templates are used to support ArcView 3 geocoders.

Modifying the file

To make changes to the template file, refer to the diagram on the following page to indicate the property name that you wish to alter within the template file. Open the template for the file type that you wish to alter and, using a text search feature of the text editing software, locate the property name. Once you have found the property name, you simply need to alter the value associated with it. Saving the file in the same folder under a different name and with a different description will simplify the process of locating the modified template when building your address locator.

Special case properties

Within the address locator file, there are several properties that you may wish to modify that require some additional explanation.

Many of the settings on the Address Locator Properties dialog box use a check box to indicate if the property is to be used. In the textual .lot file, these are defined as true—indicating that it is checked; or false—indicating that it is not checked.

The option to use relative paths is used primarily when distributing the address locator. See the chapter 'Distributing your address locator' for more information on relative paths and distributed address locators.

When indicating the valid reference data field names, the table name refers to one of several possible tables. When wanting to add another alias for the field names, you must find the correct table. By default, the table name associated with the primary reference data is referred to as 'Table1'. For example, if you wanted to add another possible field name for "street name" on the primary reference data table, the property key would be:

```
RD.Val.Table1.StreetName =
```

More information on tables will be provided in the section 'Accommodating an alternative reference data schema' in this chapter.

When adjusting the field names, you must indicate the field number. Immediately above these properties in the .lot file, you will see the fields listed. The first field listed is referred to as "field.0", followed by "field.1", "field.2", and so on.

Finding the default settings in the address locator file template

The following diagram indicates the property name within the address locator template file that corresponds to the default setting in the Address Locator Properties dialog box. For more information on what each of these settings do, see the chapter 'Building an address locator' in this book.

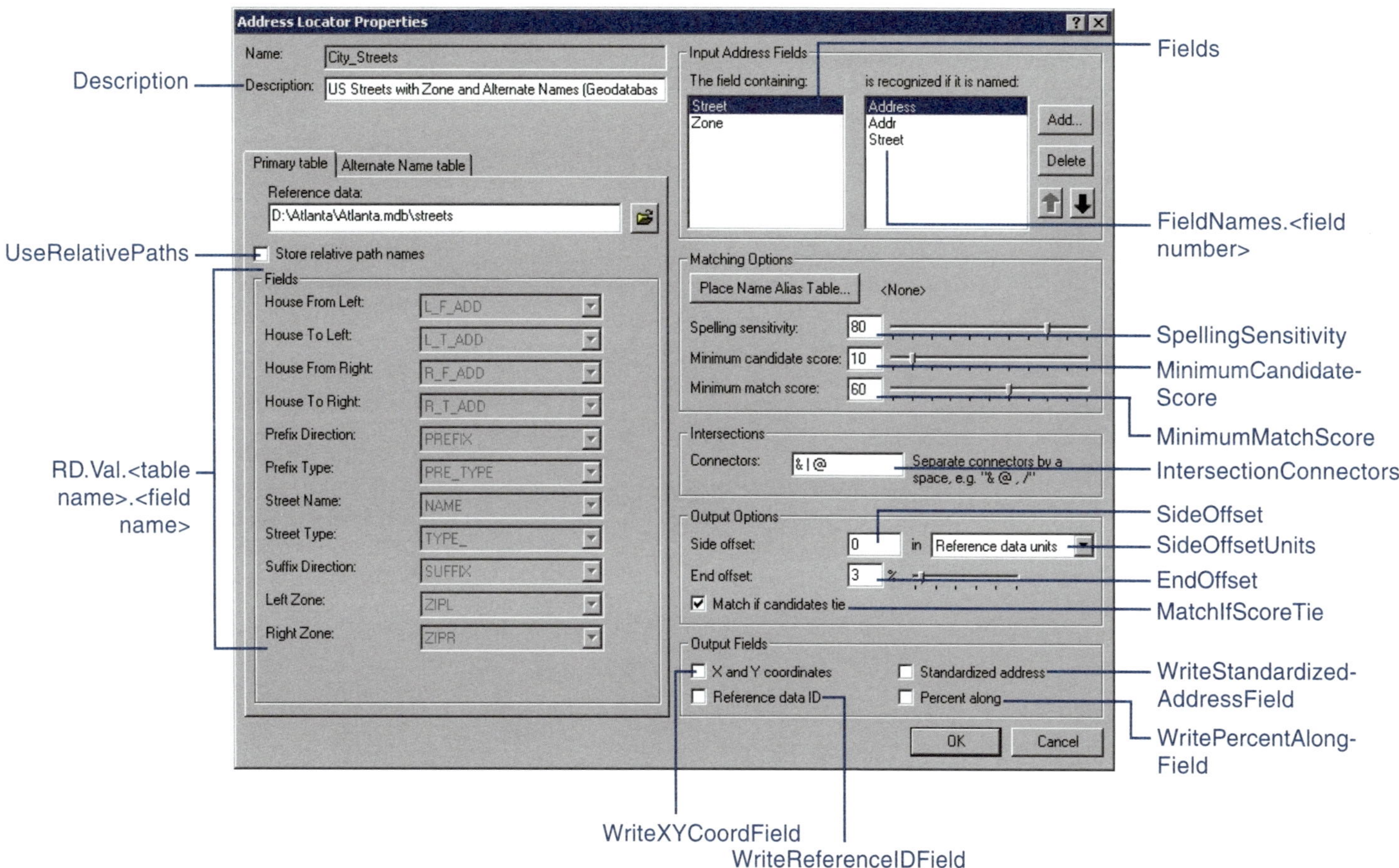

Adjusting the default settings in the address locator template file

Once you recognize the properties you wish to alter, creating a new address locator template file is a simple task. It is advisable to create a copy of a template similar to the one you wish to create, then rename and modify the copy of the template. Modifications can be made in any basic text editing software.

1. Navigate to the Locators folder within your ArcGIS install.
2. Create a copy of the .lot file that you wish to modify and rename it a recognizable name.

   This name will appear in the Create New Address Locator dialog box when creating a new address locator.
3. Open the new file with a basic text editing software. ►

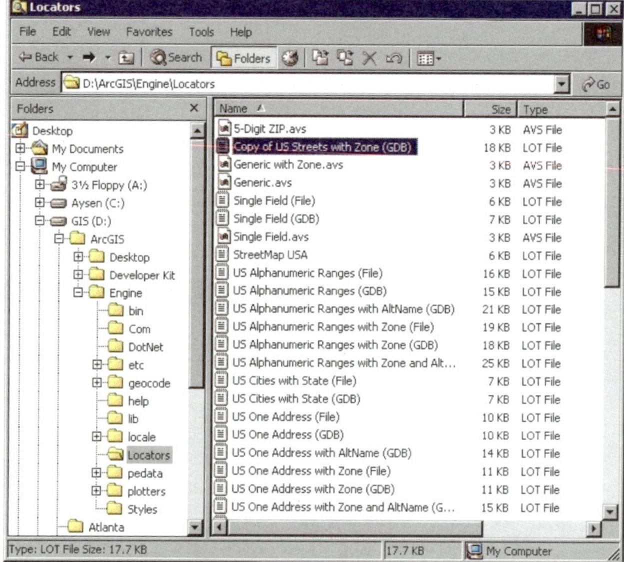

4. Using the text search function of the text editing software, search for the word “Description”.
5. Within the quotation marks, modify the description value as desired.

   This description will appear on the Address Locator Properties dialog box and in the Address Locator Manager in ArcMap. It should not exceed 255 characters.
6. Repeat steps 4 and 5 for adjusting other desired property values.
7. Save your changes and close the text editing software.

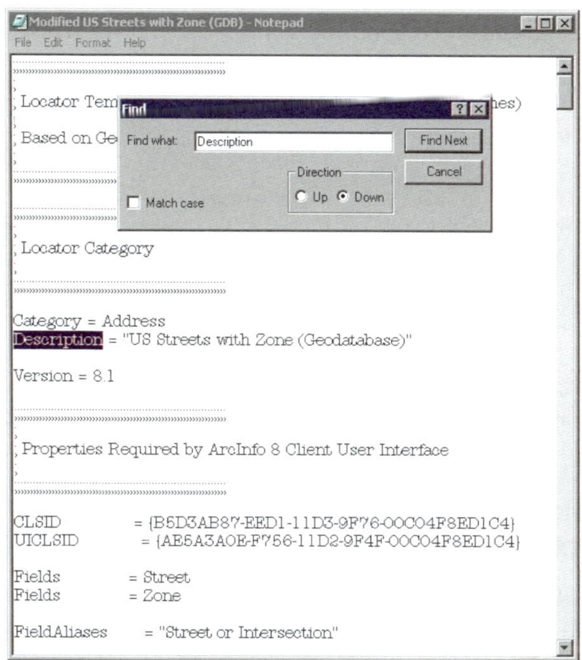

Category = Address
Description = "Modified US Streets with Zone_MyUserName"

Table schema and the locator file

Traditionally, when creating an address locator, you have one primary reference data table as well as other possible tables such as the Alternate Name table or the Place Name Alias table. While the address locator has been designed to accommodate these tables, you may wish to use a different table schema. For example, you may want to modify the address locator to accommodate a specific address data model schema, or you may have address information distributed over several tables. This section will define how these table schemas are defined within the address locator file, how the information in your attribute table is mapped to the rule base files, and finally, how you can modify the address locator to accommodate alternative table schema.

The rule base files require certain address elements

As part of the geocoding process, the .mat file within the geocoding rule base has certain address elements that it must use to search for an address location within the reference data. These elements vary based on the address locator style that you are using; however, when searching for a house number address, the address elements needed include address range information, street names, directions, prefixes and suffixes, and zone information. The .mat file must be directed to the exact field on a particular table to get this information. In order to facilitate this process, a number of steps are taken within the address locator file.

A table schema is defined

Each address locator style specifies the type of table schema that is required for the geocoding process to work. The table schema, simply put, is the number of attribute tables for the reference data and the information that will be found in those tables. For the most basic of address locators, this will include one table containing all of the required address elements. More advanced address locators, such as those that will search for alternative street names, use two or more tables.

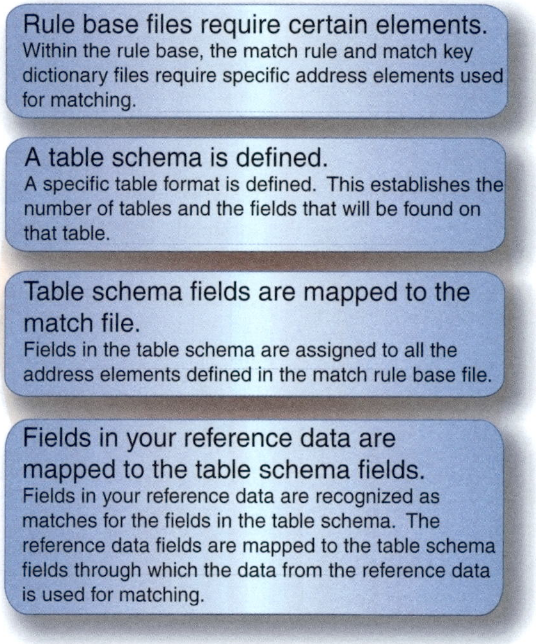

Rule base files require certain elements.
Within the rule base, the match rule and match key dictionary files require specific address elements used for matching.

A table schema is defined.
A specific table format is defined. This establishes the number of tables and the fields that will be found on that table.

Table schema fields are mapped to the match file.
Fields in the table schema are assigned to all the address elements defined in the match rule base file.

Fields in your reference data are mapped to the table schema fields.
Fields in your reference data are recognized as matches for the fields in the table schema. The reference data fields are mapped to the table schema fields through which the data from the reference data is used for matching.

The table schema is defined using several properties within the address locator file. The properties define the tables and the fields that will be found on the table:

```
RD.Table  = <table name>,"<common name>"
RD.Table = Table1,"Primary table"
```

This property establishes the basic table schema. Using the example above, one of the tables in the reference material is referred to in the table schema as "Table1". This name is used in other properties throughout the address locator file. The common name is also assigned, referring to the name used on the Address Locator Properties dialog box:

```
RD.<table name>.Field = <field name>, "<common
field name>" ,<DB indexing>,<required value>
```

```
RD.Table1.Field = StreetName,"Street Name",N,TRUE
```

Once the tables are established, fields are assigned to each of these tables. This series of properties states the type of information that should be found in each of the fields in the table schema. It is possible that your reference data will have fields that are not listed in the locator file. These fields are simply disregarded during the geocoding processes.

The first two properties are used to define the field name, determined through querying the attribute table, and a common name for the field. The DB indexing element of the value refers to the need and condition of a table created in the database. This is used when the reference data is stored within a geodatabase. A "Y" suggests that the index is created, while an "N" suggests that it is not created. In most cases, an "N" is used instead of using the database index. Instead, a geocoding-specific index is created. For more information about geocoding-specific indexes, see the subsection 'Creating index tables' later in this section. If an index is created, you can also specify if the index will be unique, descending, or both. In essence, you could define this element with "YUD" suggesting that yes, an index will be created, it is to be unique, and is to be descending. The required value element of the property value states whether a value is required for every attribute in that field. Some field values, such as prefix direction, are acceptable if left blank. However, an attribute for the street name field would be required in a search.

With these properties, the number of tables, their reference names, and the information found on these tables are established.

Table schema fields are mapped to rule base

With the table schema established, other properties are used to map the table schema fields to the rule-base-required address elements. In essence, these are telling the .mat rule base file where the information is that it requires to run the geocoding process:

```
RD.<query name>.MF = <MAT file defined name> ,
<table name>.<field name>
```

```
RD.Query1.MF = StreetName,Table1.StreetName
```

This series of properties assigns certain table schema fields to match against address elements in the .mat file. For more information on the .mat and other rule base files, see the *Geocoding Rule Base Developer Guide*.

The table and field names refer to table and field names established previously in the table-schema-related properties. This property also mentions the query name. Query information will be discussed later in this chapter.

Fields in your reference data are mapped to the table schema fields

When the table schema is established and the table schema fields are mapped to the required rule base file, the final step is mapping the fields on your attribute table to the fields in the table schema:

```
RD.Val.<table name>.<field name> = <possible
reference table field names>
```

```
RD.Val.table1.LeftTo = LEFT_TO
```

This series of properties searches through the attribute table you have provided, identifies potential field names, and maps that field to the corresponding field in the table schema. In the example provided above, the locator is searching for the field name “LEFT_TO” in your attribute table. Once found, it will map that field of your reference data to the table schema field “LeftTo”. Also, in this property, “table1” is referring to “Table1” as defined in the table schema.

In the locator file you will find a substantial list of these properties. These are used to define the potential field names within your reference data to several fields including LeftFrom, LeftTo, RightTo, RightFrom, StreetName, StreetType, and so on. Once your address locator has been created using these templates, only one reference field name with its accompanying property will be present.

Following the path of transfer from your attribute table to the table schema and finally to the rule base file might initially seem complex. However, in following this path, the address locator file is made versatile enough to accommodate a wide range of attribute tables and field names and translate them to the format that is needed for the rule base.

Once the table schema has been established, the next step includes the creation of index tables and the final attribute table query.

Creating index tables

The following material is specific to index tables created within a geodatabase. For more information regarding shapefile and coverage indexes, see the section ‘Special cases’ in this chapter.

Depending on the type of address locator you are creating, index tables might need to be created for one or more field values within your attribute table. Index tables and their use are described in detail in the chapter ‘Keeping an address locator current’ in this book. Creating and populating an index table entails the following steps:

- Define an index table.
- Establish a table name.
- Define the field properties and populate the index table with attributes from the needed fields of the attribute table.

These steps are completed through a series of index-related properties within the address locator file, the first of which defines the name for the index file used throughout the geocoding process as well as a common name for the index table:

```
RD.IdxTable  = <index table name>, "<common name>"
```

```
RD.IdxTable = IdxTable1,"IndexTable for Primary table"
```

This property establishes a formal name and a common name for the index table. Just as the table schema name was based on the table type and number (that is: ‘Table1’), the index tables are usually assigned a name such as IxdTable1, IdxTable2, and so on; however, this is left to the user’s discretion.

An index table is actually produced and stored in your database. In creating this table, an additional property defines the table name prefix that will be assigned to this new table:

```
RD.<index table name>.Prefix = <prefix name>
RD.IdxTable1.Prefix = GC_SZS
```

When the index table is created, it is assigned a name based on the prefix name value in this property as well as the object class identification. In the example shown above, the index table created would be named GC_SZS<object class ID of the source table>. The index table has been created with the assigned name.

The final step is to populate the table. This is done with the following property:

```
RD.<index table name>.IdxField = <field name>,
"<common name>", <DB Indexing>,<function and
source>
```

```
RD.IdxTable1.IdxField = Sx,"Soundex code
field",Y,@SDEX($Table1.StreetName)
```

These properties direct the creation of the index tables and the individual columns that will be created in the index table. The column name refers to the name to be used in the index table for the field, with the common name acting as a user-friendly description of the field. The DB indexing value element holds the same relevance as in the standard table, defining the need for and format of a database index based on the field. The function or source of the field values can consist of a range of information. In some instances, the information is copied directly from the reference table; for example, when creating an ID value used to join the index table to the reference table, the index value from the reference table can be copied directly over. In other cases, the data in the reference table needs to undergo some type of conversion. One example of this is the Soundex function. Below is an example of such a function:

```
@SDEX($Table1.StreetName)
```

`@`—indicates that it is a function.

`SDEX`—is the function name, as assigned in the RD.func property. These properties will be discussed later in this section.

`(...)`—is specifying the location of the data that will undergo the function.

`$Table1`—is used to specify that whatever table was used to create "Table1" should be used.

`StreetName`—specifies the table schema field name on that table to use.

Roughly translated, this is stating that the Soundex function should be run on the data in the "StreetName" field of the "Table1" table and be added as a field within the index table.

Several other functions might be used in the address locator file. These are specified with the property:

```
RD.Func = <function name>, <function library>,
<function name as declared in the library>,
"<description>"
```

```
RD.Func = SDEX,mtchloc,SX,""
```

Modifying the RD.Func properties will rarely be done and only when accommodating third party software components.

Once the attributes of the index table are populated, one additional property is used to modify the field name in the table:

```
RD.Val.<index table name>.<field name> = <user-
interface field name>
```

```
RD.Val.IdxTable1.Sx = SX
```

This property defines a name to be used when the index table is opened in ArcMap or viewed in ArcCatalog. While not necessary, you may wish to use a different, more user-friendly name when displaying the table, yet preserve the field name for functionality in the locator file. In several database systems, these field names must also consist of all capital letters. In the example provided above, this would be "SX".

Understanding the query process

The query process is rather straightforward in the locator file. In short, a query name is defined, and a series of table and field values are matched to the related fields defined in the rule base to use as search parameters:

```
RD.Query = <query name>, "<common query name>"
```

```
RD.Query = Query1,"Main query"
```

Just as tables were often assigned the name "Table" followed by the table number, queries are also assigned names such as "query1" and "query2". The common name used to describe the query is at the user's discretion.

Select condition query properties

In addition to these query properties, you will also find some special case query properties:

```
RD.<query name>.Where = <expression>
```

```
RD.Query1.Where = $IdxTable1.Sx = '@SDEX(%SN%)'
```

Often, there will be certain functions that need to be performed on the input address data before it can be compared to the values in the reference data. For example, when calculating street names, an index value has been created within the reference data. The same index value must be calculated for the input address data.

The example provided above illustrates this type of property. It states that in order to compare with the values in the index table, the value of each address element after the input address is parsed and standardized, referred to as the match key value (as defined in the match key dictionary rule base file), and must also undergo the Soundex function.

In addition to the values related to the "where" query property, there are several other properties that are tied with the "where" value statement. These have been divided and assigned their own property name; however, they are concatenated when read by the geocoding engine:

```
RD.<query name>.InvariantWhere = " AND $<table
name>.<field name> = $<table name>.<field name>"
```

```
RD.Query1.InvariantWhere = " AND $Table1.ID =
$IdxTable1.XID"
```

The InvariantWhere properties simply define how the tables are joined together. In the example provided above, the ID value from "Table1" is used to link to the "XID" values on "IdxTable1":

```
RD.<query name>.ConditionalWhere = <match key
token>, "expression"
```

```
RD.Query1.ConditionalWhere  = ZN, "AND
($IdxTable1.Zone = '@ZNSX(%ZN%)')"
```

The ConditionalWhere property states additional properties to add to the "where" condition, but only if certain conditions are met. The example provided above illustrates this property.

This example states that if the "ZN" match key token (the zone) contains a nonnull value within the input address data, then the search will add another restriction by appending the ConditionalWhere expression to the previous Query1 where expression. For example, if an address used for geocoding contains a value "92373" in the zone field, the geocoding engine will search for the possible matches using both street name and zone value. If the zone field is blank, the geocoding engine will search for matches using the street name only, as the search restriction is relaxed. As a result, it is recommended not to use '0' (zero) in the address table for the ZIP Code field. Use a blank space instead. You can define multiple ConditionalWhere expressions if applicable.

Accommodating an alternative reference data schema

Now that you understand how the address locator file communicates the information in the reference material to the rule base files, you will learn how to accommodate alternative reference data table schema.

Why use an alternative table schema?

As described in the previous section, the address locator is designed to accommodate a specific table schema. This table schema usually entails a primary reference data table and possibly a Place Name Alias table or an Alternate Street Name table. For most basic operations, this table schema accommodates the material that you have available.

However, you might have your spatial information stored in an alternative format. One example of this would be to have street and address information stored separately in a database. That is to say that you might have the street names in one table, the segment address ranges in another, and zone information, such as city or postal code, in another.

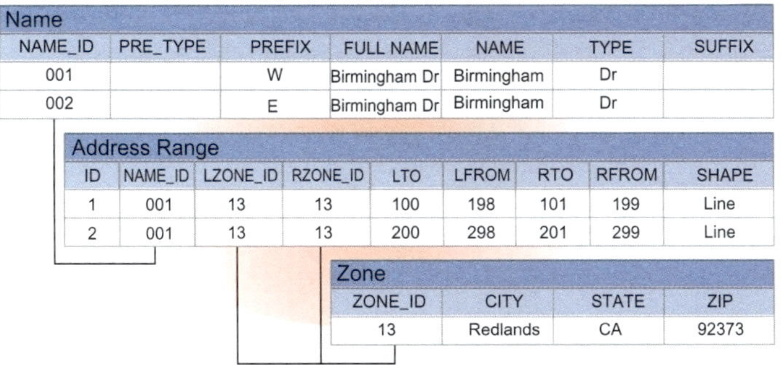

Name

NAME_ID	PRE_TYPE	PREFIX	FULL NAME	NAME	TYPE	SUFFIX
001		W	Birmingham Dr	Birmingham	Dr	
002		E	Birmingham Dr	Birmingham	Dr	

Address Range

ID	NAME_ID	LZONE_ID	RZONE_ID	LTO	LFROM	RTO	RFROM	SHAPE
1	001	13	13	100	198	101	199	Line
2	001	13	13	200	298	201	299	Line

Zone

ZONE_ID	CITY	STATE	ZIP
13	Redlands	CA	92373

Instead of all of the address information being stored in one table, you may find that your address information is distributed across several tables. In this diagram, all of the tables contain different address elements, yet they are joined together with their corresponding ID field. This example will be used throughout this section in describing how to accommodate alternative table schemas.

This table schema could have a variety of advantages such as in editing and incorporating multiple geometries related within the database. For more information on this type of data storage, see *Designing Geodatabases*.

The one drawback to this type of data storage is that a typical address locator file will not accommodate your reference data table schema. You will need to either create or modify an address locator file in order to function with your specific schema. In order to do so, several properties need modification within the locator file.

One thing to keep in mind through this process is that you are simply redefining the properties described in the previous section. The rule base file still requires the same address elements; the only issue that you need to correct is that your information is in several tables instead of one. The processes needed are still the same.

In the following paragraphs, the modification process will be described in general terms, with specific examples relating to the table schema diagram previously illustrated on this page.

Redefining the table schema

As described in the previous section, when defining the table schema, you are, in essence, expressing the format that the data is stored in; you are describing what fields are in what table in the reference data. This is done by defining the tables that will be present and by stating what fields will be found in those tables.

Instead of specifying one table, this table schema consists of three tables to be used. In the previous section, you learned two properties that were used to define the table schema. These were the RD.Table and RD.<table name>.Field properties. To accommodate the table schema illustrated on the previous page, these properties would need to be added, similar to the following:

```
RD.Table = name,"Name table"
RD.Table = addr,"Address table"
RD.Table = zone,"Zone table"
```

These properties define the tables that will be present. You may note that in this example, the table names were not assigned the name “Table1”, “Table2”, and so on. Instead, names related to their table properties were used. This name is left to your discretion; however, you will need to be sure that all of the other properties refer to these table names consistently.

The following properties will establish which table contains the needed address elements. When modifying these properties, you might find it easier to simply change the table name portion of the property, leaving all of the value elements as they are. For example, the following property value is present on the US streets with zone address locator template file:

```
RD.Table1.Column= LeftFrom,"House From
Left",N,TRUE
```

To modify this property, you simply need to change the table name from “Table1” to “addr” as such:

```
RD.addr.Column= LeftFrom,"House From Left",N,TRUE
```

“addr” refers to the table name you established in the previous RD.Table properties. Once complete, this section would appear as follows:

```
RD.addr.Column= LeftFrom,"House From Left",N,TRUE
RD.addr.Column= LeftTo,"House To Left",N,TRUE
RD.addr.Column= RightFrom,"House From
Right",N,TRUE
RD.addr.Column= RightTo,"House To Right",N,TRUE
RD.addr.Column= ID,"Record ID for the address
table",N,TRUE
RD.addr.Column= NameID,"Reference ID for name
table",N,TRUE
RD.addr.Column= ZoneID,"Reference ID for zone
table",N,TRUE
RD.addr.Column= Shape,"Shape field",N,TRUE
RD.name.Column= PreDir,"Prefix Direction",N,FALSE
RD.name.Column= PreType,"Prefix Type",N,FALSE
RD.name.Column= StreetName,"Street Name",N,TRUE
RD.name.Column= StreetType,"Street Type",N,FALSE
RD.name.Column= SufDir,"Suffix Direction",N,FALSE
RD.name.Column= ID,"Record ID for the name
table",N,TRUE
RD.zone.Column= LeftZone,"Left Zone",N,TRUE
RD.zone.Column= RightZone,"Right Zone",N,TRUE
RD.zone.Column= ID,"Record ID for the zone
table",N,TRUE
```

Notice that the only parts of these properties that have changed from the original are the table names, which have been altered to accommodate the table names specified previously. Also, additional columns have been added to link the tables together. Leaving the value elements unaltered simplifies the process and places the same guidelines on the field properties.

The table schema is now established. The next step is to map the table schema fields to the rule base.

Mapping table schema fields to the rule base

The `RD.<query name>.MF` is used to map the table schema fields to the appropriate .mat file requirements within the rule base files in the geocode folder. Again, this is simply a process of slightly modifying the table names. For example, the US streets with zone address locator template file contains the following:

```
RD.Query1.MF = LeftFrom,Table1.LeftFrom
```

You might immediately notice the reference in the value to "Table1". That table name has been altered, so that is one part that needs to be changed. If you refer to the previous section, you will see that the other value elements represent the defined .mat filename (this does not change) and the field name specified in the table schema. You have not changed this value either. With this in mind, the only parts of these properties to change are the table names to which they refer. You will find it easier to determine the table names by referring to the table schema field list you modified previously. Using the US streets with zone example again, the list of properties would appear as follows. The table names have been placed in bold typeface to facilitate their recognition:

```
RD.Query1.MF = LeftFrom,addr.LeftFrom
RD.Query1.MF = LeftTo,addr.LeftTo
RD.Query1.MF = RightFrom,addr.RightFrom
RD.Query1.MF = RightTo,addr.RightTo
RD.Query1.MF = PreDir,name.PreDir
RD.Query1.MF = PreType,name.PreType
RD.Query1.MF = StreetName,name.StreetName
RD.Query1.MF = StreetType,name.StreetType
RD.Query1.MF = SufDir,name.SufDir
RD.Query1.MF = LeftZone,zone.LeftZone
RD.Query1.MF = RightZone,zone.RightZone
RD.Query1.MF = Shape,addr.Shape
RD.Query1.MF = ID,addr.ID
```

Mapping the fields in your reference data to the table schema fields

You have already changed the properties in the address locator template file to recognize a different table schema, and you have mapped those table fields to the correct requirements in the rule base files. The final step is to find the actual fields in your reference data to map to the table schema fields.

In the previous section, a substantial list of RD.Val properties was given. In this list, the address locator searches your attribute table for the field that will match with the table schema field. These appear as follows:

```
RD.Val.Table1.LeftFrom = L_F_ADD
```

In common terms, this property is stating that a recognized field name on your reference data that corresponds to the "LeftFrom" field on "Table1" would be "L_F_ADD". For each field on the table schema, a long list of potential reference data field names are listed. In determining what needs modification in these properties, you can examine the previous example.

You still wish to search for a valid value, so it would be best to not change the "RD.Val" portion of the statement. You will note that there is another reference to "Table1", which you have changed into three different table names including "name", "addr", and "zone". The "LeftFrom" refers to the table schema field name, which you did not alter. The potential reference data field name should not be altered either. In essence, again, all you need to modify in these properties is the table name, based on the table that contains the particular field. In the above example, you know that the "LeftFrom" address element is found on the "addr" table. You would modify this property to the following:

```
RD.Val.addr.LeftFrom = L_F_ADD
```

Once this is complete, the index tables need to be generated for some of the fields, and some modifications are needed in the query process to indicate the variety of tables as opposed to the individual table.

You may notice in the three-table schema that the zone table contains both city and postal code information. In most of the default address locator files, the zone field will be mapped to the postal code or city. The address type used is based on the order of the potential field names within this list of properties. Once a match is found, the remainder will be disregarded.

Alternatively, if you wish to look only for the postal code, you may wish to remove some of the properties that search for city information. For example, in the address locator you are modifying, you may wish to remove the following:

```
RD.Val.Table1.LeftZone = CITY
RD.Val.Table1.RightZone = CITY
```

In doing so, the address locator will not recognize the field "CITY" in the reference data as the zonal information, but it will find the "ZONE" field. For more information regarding searching for multiple zone fields, see the *Geocoding Rule Base Developer Guide*.

Creating index tables

As stated in the previous section, creating and populating an index table entails the following steps:

- Define an index table.
- Establish a table name.
- Define the field properties and populate the index table with attributes from the needed fields of the attribute table.

In the table schema that is being accommodated, the address elements that require indexing are distributed across two tables, a street name index for the name table and a zone index for the zone table. To accommodate this, two index tables will need to be established. The `RD.IdxTable` properties are used to define the index tables. In these properties, the index table name is typically referred to as "IdxTable1" and so forth, as shown in the following example:

```
RD.IdxTable = IdxTable1,"IndexTable for Primary table"
```

You may wish to leave the table name, or you can modify it to be more intuitive, such as "NameIndex" instead of "IdxTable1". You may also wish to modify the common name to refer to the table. Neither of these modifications is required, but they may help simplify other modifications you will make.

You may also wish to modify the table prefix value. The prefix should consist of all capital letters, be less than seven characters long, and contain no spaces. If you modify this, it will make recognition easier when the table has been created and stored within your geodatabase.

In populating the table, you will need to redirect the table names to the corresponding tables, which, by default, will refer to "Table1" and so forth. Each table name is accompanied by the field name, so recognizing the table where the information is to be found should not be difficult. The indexing section from the US streets with zone address locator template file would be modified as follows (the changes have been put into bold typeface):

```
RD.IdxTable = NameIndex,"Index Table for name table"
RD.NameIndex.Prefix = NAM_IDX
RD.NameIndex.IdxField = XID,"Soundex foreign key field",Y,$name.ID
RD.NameIndex.IdxField = Sx,"Soundex code field",Y,@SDEX($name.StreetName)
RD.Val.NameIndex.XID = XID
RD.Val.NameIndex.Sx = SX
```

This established the index table for the name table.

```
RD.IdxTable = ZoneIndex,"Index Table for zone table"
```

```
RD.IdxTable1.Prefix = ZON_IDX
```

```
RD.ZoneIndex.IdxField = XID,"Index foreign key field",Y,$zone.ID
```

```
RD.ZoneIndex.IdxField = ZoneL,"Left zone field",Y,@ZNSX($zone.LeftZone)
```

```
RD.ZoneIndex.IdxField = ZoneR,"Right zone field",Y,@ZNSX($zone.RightZone)
```

```
RD.Val.ZoneIndex.ZoneL = LZONE
```

```
RD.Val.ZoneIndex.ZoneR = RZONE
```

This establishes the index table for the zone table. These modifications will create the two index tables accommodating the alternate table schema.

Redefining the query process

The query process refers to the tables and table schema; these properties will also need to be modified.

The `RD.<query name>.Where` properties state how the address data being searched should be processed in order to match the indexed values found on the corresponding index tables.

Throughout all of the query properties, you will see specific schema table names or the index table names and fields names referred to together, as shown in the property:

```
RD.Query1.Where = $IdxTable1.Sx = '@SDEX(%SN%)'
```

However, you have created two index tables and have changed the index table name, so this property will need to be changed to the following:

```
RD.Query1.Where = $NameIndex.Sx = '@SDEX(%SN%)'
```

Note that only the index table name was changed.

The `RD.<query name>.InvariantWhere` property is used to link tables together in such a way that the query remains valid even when the ConditionalWhere query does not apply. Typically, this is only used to link the primary table and the index table; however, in the table schema, three tables are being used and, thus, will need to be joined via the InvariantWhere property value sequence. As you can see from the example, this property defines the field on one table and its corresponding field on another table, such as linking the "ID" field in "Table1" to the "XID" field on "IdxTable1":

```
RD.Query1.InvariantWhere = " AND $Table1.ID = $IdxTable1.XID"
```

Some of the table names have changed and other table links need to be made. Using this alternative table schema example, this would look like the following:

```
RD.Query1.InvariantWhere=
" AND $addr.NameID = $name.NameID
AND $addr.ZoneID = $zone.ZoneID
AND $name.NameID = $nameIndex.XID"
```

You will see that each of the links have been defined, including the link of the index table with the address table, the address table with the name table, and the address table with the zone table.

To accommodate the index table created for the zone field, you may wish to use the ConditionalWhere property. The additional condition suggests that this procedure only needs to be done when, in this case, a zone property for the input address data contains a non-null value. The ConditionalWhere property would be as follows:

```
RD.Query1.ConditionalWhere =ZN, " AND
$zone.ZoneID = $zoneIndex.XID AND
("$ZoneIndex.ZoneL = '@ZNSX(%ZN%)'OR
$ZoneIndex.ZoneR = '@ZNSX(%ZN%)')"
```

Intersection queries

Thus far in this section you have learned the methods of modifying the query for address queries. The intersection query properties are closely related; however, they are a distinct group of properties and will also need to be modified. Use the same principles defined for an address query to modify these properties.

Modifying the query

There are a number of reasons that you may wish to modify your query. In the previous section, you already have modified the query process, directing the query properties to recognize the alternative table schema that you established. Beyond modifying the query to accommodate any changes to the table schema, there are several other reasons that you may wish to use an alternative table schema, such as when using an Alternate Name table. Several address locator template files have been provided that already have this alternative query process established. In these locator files, the additional table is established in the table schema; then, a second query is developed. While the first query looks for the address using the name provided in the primary reference data, the second query looks for the address but uses the name field from the Alternate Name table. There are still other reasons that you may wish to add to or modify the query. For example, you may wish to create one address locator that will query two different feature classes, such as when using two feature classes with different table schemas.

In this section, another type of modification to the query will be described. You will learn how to modify the query to accommodate a table schema in which the alternate names were included as an additional column on a standard attribute table.

Reviewing the data mapping process

Previously in this chapter you learned about the table schema and how it relates to the query process. These same concepts will be used in understanding the modification possibilities within the address locator file.

Rule base files require certain elements.
Within the rule base, the match rule and match key dictionary files require specific address elements used for matching.

A table schema is defined.
A specific table format is defined. This establishes the number of tables and the fields that will be found on that table.

Table schema fields are mapped to the match file.
Fields in the table schema are assigned to all the address elements defined in the match rule base file.

Fields in your reference data are mapped to the table schema fields.
Fields in your reference data are recognized as matches for the fields in the table schema. The reference data fields are mapped to the table schema fields through which the data from the reference data is used for matching.

While many things can be altered or modified within the address locator file, the geocoding rule base still has certain address elements that it needs to map corresponding match fields. While this does not change, you may wish to send alternate address elements to these match fields. For example, when using an Alternate Names table, you first send the name from the primary reference data table, then it will perform a second query and send the alternative name. In the example that will be explained in this section, first the values from the name column will be delivered, then, in a second query, the values from the alternate name column will be queried.

Defining the table schema

When defining the table schema, you simply need to direct the table schema to find another field on the specific table. In the case of one additional field on the attribute table, this would be done by adding the following property:

```
RD.Table1.Field = AltStreetName,"Alternate Street Name",N,FALSE
```

Mapping table schema fields to the rule base

The goal in attempting to accommodate the alternative name column is to first search for the name information, then search for the alternative name information. To do this, you will need to create a second query. The first query will send a specific group of address elements to the rule base. The second query will send a different group of elements to the rule base. The initial query has already been established within the address locator file. The second query is similar, with a few changes. First, the query name is no longer "Query1" but is named something different, such as "Query2". Again, the name used is at your discretion; for example, you could name the first query "Name_Query" and the second "AltName_Query", indicating that the first is to search using the name values and the second is to search using the alternate name values.

In addition to changing the query name, you must direct the query to send the alternate name values to the rule base, as opposed to the standard name. This is done by changing the field name for the name query properties. For example, instead of:

```
RD.Query1.MF = StreetName,Table1.StreetName
```

you would change it to:

```
RD.Query2.MF = StreetName,Table1.AltStreetName
```

Note that the query name has been changed, and the "AltName" column is being indicated.

In review, mapping the table schema fields to the rule base consists of the following steps:

- Copy the RD.<query name>.MF properties that you developed in the previous section.
- Paste the copy below the first set of RD.<query name>.MF properties.
- Change the name of the query in each of the pasted properties.
- Change the attribute-specific properties to refer to the new attributes.

Mapping reference data fields to the table schema fields

The next step is to map the additional reference field to the table schema fields. Within the `RD.Val.<table name>.<column name>`, you will need to add information for the Alternate Name table schema field and the recognized field name for the table schema. This would be as follows:

```
RD.Val.Table1.AltName = ALT_NAME
```

You may also wish to add other potential field names for the alternate name field in the reference data table.

Adding an additional column to the index table

Just as your standard street names column has a corresponding index table column, you will also want to create an index column for your alternate name values.

This additional column would be generated using the Soundex function and would be added to the "IndexTable1" table. The table is already defined; the new index column will be added to the index table. This would be done by adding the following properties with the other properties referring to the index table:

```
RD.IdxTable1.IdxColumn = Alt_Sx,"Soundex code
column for alternative name",Y,@SDEX
($Table1.AltName)
```

This property will create and populate an additional field in the index. The final step is in assigning the field a name. This is done in the following manner:

```
RD.Val.IdxTable1.Alt_Sx = ALT_SX
```

Redefining the query process

In adding the query process properties, you may wish simply to copy the default query process properties for "Query1" and paste them in place for the "Query2" and make minor modifications to accommodate the additional query.

Included with the "Query1" query process definition properties, you will find the following property:

```
RD.Query1.Where = $IdxTable1.Sx = '@SDEX(%SN%)'
```

In review, this basically states that the values in the "Sx" column on the "IdxTable1" were processed using the Soundex function, and when searching against them, you must process the input data the same way. To accommodate the "Alt_sx" column, the following changes should be made:

```
RD.Query2.Where = $IdxTable1.Alt_Sx ='@SDEX(%SN%)'
```

You will note that the only modifications made were the change in query name and the reference to the "Alt_Sx" column in the index table.

While the example described in this section might not cover your exact procedures, as they vary based on the specific table schema and application, the concepts presented here will help you in understanding the changes you need to make to suit your specific need.

Accommodating changes in the rule base files

In a traditional address locator there are certain search parameters that the Find dialog box searches or that are extracted from a table of addresses when batch geocoding. While a wide range of address locators are provided, you may find that these address locator file templates do not accommodate the address elements that you wish to use in the search. One example of this might be if you wanted to search for a location based not only on the "Zone" field, such as the postal code, but instead using several zonal categories including the city and postal code as well as the state or province.

Determining your goal

Before beginning the process of modifying the rule base files, it is important to understand clearly what your final goal is and if your address and reference data can accommodate these modifications.

What additional address elements do you wish to accommodate?

The most common reason that one changes the rule base files is to accommodate additional address elements in the search. You may find that there are one, two, or more address elements that the default address locator styles do not accommodate that you wish to use in your search. One example of this is in accommodating both the city and the postal code in a search. By clearly defining your goal, the steps and procedures will be clarified, and your efforts will be focused on accommodating the specific goals.

Does your data support these additional address properties?

In order to accommodate alternative address elements, it is also important to note if your address and reference data contain the additional information that you would like to search. For example, if you wanted to search for your addresses using both the city and postal code information, your reference data attribute table would need to have a field that contains the postal code and a field that contains the city name. If a good deal of manipulation or editing is required to accommodate this, you might consider the effort required to do this against the advantages related to a search that accommodates the two zone fields.

Modifying your match file

Before you make any changes to the rule base files, it is suggested that you learn more about their functionality and their related properties. This information is provided in the *Geocoding Rule Base Developer Guide*. In this guide you will find a vast amount of information related to the match file, with the extension .mat; the match key dictionary, with the extension .dct; and all of the other rule base files. To obtain a copy of the Geocoding Developer Kit including the *Geocoding Rule Base Developer Guide*, contact the customer service department at ESRI. It is also recommended that before modifying these files, you make a copy and rename the file to a name of your choosing. These files are used for a wide range of address locators and, at times, other applications. You should modify only the copy of the file that you created.

Within the match file there is a list of variable properties and match properties related to each of the address elements that are to be queried through the geocoding process. These relate to the address elements that are found in the address data. One example of a variable property is as follows:

```
VAR StreetName 47 28  S ; Street name
```

The corresponding match property would look like this:

```
MATCH UNCERT  SN StreetName 0.9 0.01 700.0
```

Details about the specific elements of these properties are described in the *Geocoding Rule Base Developer Guide*.

When adding search fields, similar properties must be created or modified to accommodate the new fields. For example, to modify the single zone field to search for two zone fields, such as postal code and city, you would begin with:

```
VAR Zone 53 20 S ; Zone
```

and convert it to this:

```
VAR City 57 40 X ; City
VAR ZIP 117 10 X ; ZIP
```

Related modifications also need to be made to the MATCH variables within the match file. These might appear as such:

```
MATCH UNCERT  CT  City       0.8 0.1 700.0
MATCH UNCERT  ZP  ZIP        0.8 0.1 700.0
```

Modifying your match key dictionary to accommodate changes

Within the MATCH variables in the match file, the third element defines a two-letter match key that is used within the rule base to specify the specific address element. When you modify the match file, corresponding changes also need to be made to the match key dictionary (extension .dct).

Note that the .dct file is associated with a set of standardization files in the folder. You may not want to rename the file. However, you can edit the existing file such as the us_addr.dct file.

When searching for a single zone field, the corresponding match key dictionary variable appears as follows:

```
ZN   C    20   S; Zone
```

The "ZN" is the match key that was previously assigned to the zone address element. A thorough description of the other variable elements is described in the *Geocoding Rule Base Developer Guide*. In order to accommodate the changes mentioned in the previous section, two new match keys can be added to the file. You may not want to remove the "ZN" key if other address locators have referred to it:

```
CT   C    40   S; City
ZP   C    10   S; ZIP
```

Modifying the locator file to accommodate changes

Once your rule base files have been modified to accommodate the additional address elements, there are several modifications that need to be done within the address locator template file to accommodate these changes and the desired results. Many of these steps are repeated from other sections within this chapter.

Adjusting the search field properties

One of the first groups of properties that you will find within the address locator file refers to the search fields. The field properties are laid out in a specific format. You will note that the first subgroup contains a specific number of properties. The same number of corresponding properties exists for the other field-related properties. For example, in a number of address locator files, you will find two "Fields" properties. Under these, you will find two "FieldAliases" properties. The first "FieldAliases" property is related to the first listed "Fields" property. As you change the first set of "Fields" properties, all of the field-related properties will need to be modified accordingly.

The first in this group is the "Fields" properties. These properties define the internal name to be used for the search query. Most address locators search based on the "Street or Intersection" and the "Zone". The Field properties that correspond to these are:

```
Fields = Street
Fields = Zone
```

To accommodate the modifications that have been described in the rule base files, the zone field has been deleted, and fields for the city and postal code have been added. This would change the series of "Fields" properties to the following:

```
Fields = Street
Fields = City
Fields = ZIP
```

Closely related to the "Fields" properties are the "FieldAliases" properties. These specify the labels that are used within the Find dialog for each text box. In the graphic provided below, the "FieldAlias" properties would be as follows:

```
FieldAliases = "Street or Intersection"
FieldAliases = "Zone"
```

You will find these values used on the Find dialog box.

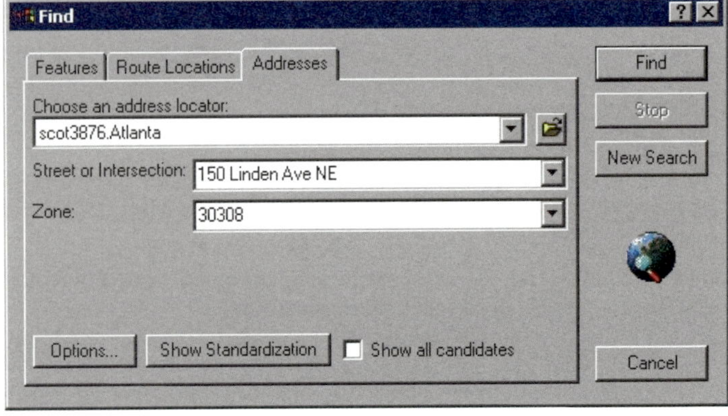

The FieldAlias properties within the address locator file are used in the Find dialog box to label the corresponding text boxes.

In order to change your query options, you will need to modify these properties. In order to search using the city and postal code field, you may wish to delete the reference to the zone field and add the following "FieldAlias" properties:

```
FieldAliases = "City"
FieldAliases = "ZIP Code"
```

These changes will modify the Address tab on the Find dialog box, providing a text box in which to enter the address, the city, and the ZIP Code.

The next property within the address locator file related to the field properties is the "FieldNames" properties. These are used within the address locator when searching for a table of addresses. In the majority of address locators that search for address and zonal information, these properties appear as follows:

```
FieldNames.0 = Address
FieldNames.0 = Addr
FieldNames.0 = Street
FieldNames.1 = Zip
FieldNames.1 = Zipcode
FieldNames.1 = City
FieldNames.1 = Zone
```

As you can see, the "FieldNames.0" key refers to the address field, while the "FieldNames.1" refers to the zone information.

You will need to modify these in order to search for the city address element and the postal code address element individually. In specifying values, you will need to examine the address table you wish to search and specify field names used. For example, if your address table had the field name "City" for the field that contained the city names and "Zip" for the column that contained the postal code information, you would modify the above fields in this way:

```
FieldNames.0 = Address
FieldNames.0 = Addr
FieldNames.0 = Street
FieldNames.1 = City
FieldNames.2 = Zip
```

You may wish to specify other potential field names in order to accommodate a range of address tables to be searched.

You may also note a few remaining field-related properties. These include "FieldRequired" and "FieldSizes". The FieldRequired property specifies whether the field is required, indicated with a value of "TRUE" if it is required and, "FALSE" if it is not required. If you are unsure if your address data contains both the city and postal code information, you may wish to assign these properties a value of "FALSE."

The "FieldSizes" property indicates the size of the text box to appear on the Find dialog box and the acceptable length of the field. The numeric value provided as its value indicates the number of acceptable alphanumeric symbols.

Specifying the match key properties

Within the address locator, the address elements for the address being searched are assigned to specific match keys that are defined within the match file. Using the previous default example where only the address and zone are being queried, these would appear as follows:

```
MKeyField = XX
MKeyField = ZN
```

The series of these properties also corresponds to the "Fields" properties explained earlier. In other words, the first property corresponds to the address field and the second corresponds to the zone field.

The XX match key value specified in these properties indicates that the information that is within this field will need to undergo some standardization process. In short, this standardization will take the address and break it up into its corresponding address elements. Further information on the standardization process can be found in the *Geocoding Rule Base Developer Guide*.

The second MKeyField listed in this example will assign the zone information to the ZN match key.

In the example modification of the match file, the ZN match key has been eliminated and two new match keys have been created. These are "CT" for "City" and "ZP" for "ZIP". In order to accommodate these new match keys, the MKeyField properties would appear as follows:

```
MKeyField = XX
MKeyField = CT
MKeyField = ZP
```

This will continue to standardize the address field as well as direct the city field to the CT match key and the postal code to the ZP match key.

Specifying the match file to be used

The address locator specifies the match file to be used in processing the address type. If you wish to change this property to use the modified match file you created, this is done through the "FileMAT" property. Your modification would replace the preexisting property with the following:

```
FileMAT = <your specified file name>.mat
```

Providing match key alias names

One of the final match-key-related properties within the address locator is the "MKeyAlias" property. This series of properties provides a more user-friendly name to the standardization results as they relate to the specific match key values. In order to accommodate the modified match file, the alias names will need to be assigned to the new match keys. For example, in order to accommodate the zone match key, the following property exists:

```
MKeyAlias.ZN = Zone
```

This will be replaced with the following two properties:

```
MKeyAlias.CT = City
MKeyAlias.ZP = Zip
```

Creating a new reference table schema

So far in this section, you have learned how to modify the rule base files and have learned the modifications required to accommodate the address data. The following information discusses modifications necessary to accommodate the reference data. In essence, this is the same procedure outlined in the section 'Accommodating an alternative reference data schema' in this chapter. In that section, the address locator was being modified to consider three related tables as opposed to one table. In this example, you are using a single table; however, instead of lumping the zonal information into one field, the table schema must consider the postal code field as well as the city field. You may wish to review the material provided in the sections 'Table schema and the locator file' and 'Accommodating an alternative reference data schema' in this chapter.

Redefining the table schema

In most scenarios, the table schema searches for only one zonal type. For example, the address locator might specify zone-related address elements that refer to the postal code or city. In order to accommodate the changes to the rule base file presented previously in this section, the table schema needs to be modified to account for both postal code and city. The addition and modification of columns within the table schema are done through the RD.<table name>.Column properties. Within these properties, you will find each of the columns that are to be found within the reference data table. 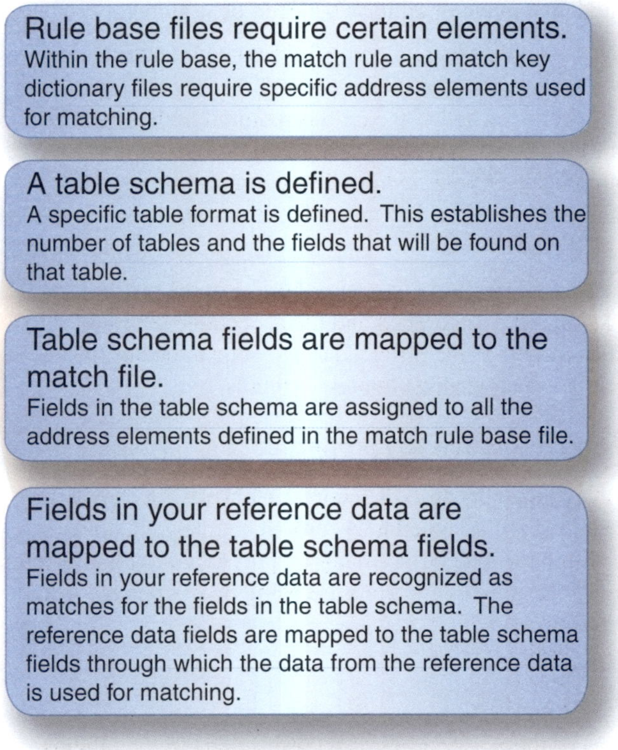

Rule base files require certain elements.
Within the rule base, the match rule and match key dictionary files require specific address elements used for matching.

A table schema is defined.
A specific table format is defined. This establishes the number of tables and the fields that will be found on that table.

Table schema fields are mapped to the match file.
Fields in the table schema are assigned to all the address elements defined in the match rule base file.

Fields in your reference data are mapped to the table schema fields.
Fields in your reference data are recognized as matches for the fields in the table schema. The reference data fields are mapped to the table schema fields through which the data from the reference data is used for matching.

As discussed previously, the table schema is changed in order to accommodate the additional zonal fields.

In order to accommodate the change, you will need to modify the property referring to "Zone" and add properties indicating both postal code and city information. In this locator file, the following zone-related `RD.<Table Name>.Column` property is found:

```
RD.Table1.Column = Zone,"Zone",N,TRUE
```

This property considers one zone field. To remedy this, the following properties should be used to replace the previous:

```
RD.Table1.Column = City,"City",N,TRUE
RD.Table1.Column = Zip,"Postal Code",N,TRUE
```

This will modify the table schema to consider a column for city information and a column for postal code information.

Mapping table schema fields to the rule base

The `RD.<query name>.MF` properties are used to assign a field within the table schema to the specific match fields within the match file.

For example, the following property assigns the zone column from the table schema to the "Zone" match field within the match file.

```
RD.Query1.MF = Zone,Table1.Zone
```

In order to accommodate the changes made to the match file, it is important to recall the changes that were made. The zone match key was eliminated and was replaced with the "City" and "Zip" match fields. These modifications would appear as such:

```
RD.Query1.MF = City,Table1.City
```

Roughly translated, this states that the "City" match field within the match file should be referenced to the table named "Table1" and the column "City" within the table schema. To accommodate the postal code information, the following property would also be added:

```
RD.Query1.MF = Zip,Table1.Zip
```

Mapping the fields in your reference data to the table schema fields

The `RD.Val.<Table Name>.<Column Name>` properties are used to map potential column names in your reference data to the table schema columns. An example of the zone field in a traditional address locator file might appear as follows:

```
RD.Val.Table1.Zone = ZIP
RD.Val.Table1.Zone = ZIPCODE
RD.Val.Table1.Zone = ZONE
RD.Val.Table1.Zone = CITY
```

In this example of changes made to the rule base files, the "Zone" column name was replaced with two columns including "City" and "ZIP". To accommodate this change, you may wish to specify only the column name on the reference data you wish to accommodate, or you may wish to add other potential column names in the event that the address locator is used with more than one specific set of reference material. This example would appear as follows:

```
RD.Val.Table1.City = CITY
RD.Val.Table1.City = CTY
RD.Val.Table1.City = CITY_ZONE
RD.Val.Table1.Zip = ZIP
RD.Val.Table1.Zip = POST_ZONE
RD.Val.Table1.Zip = ZIPCODE
```

Creating additional index fields

The default setting for creating an index value for the zonal information is to use the ZNSX function to process a numeric postal-type code and produce a standardized numeric value that can be queried more quickly. This is defined using the

`RD.<index table name>.IdxColumn` property and looks as follows in the majority of address locator files:

```
RD.IdxTable1.IdxColumn = Zone,"Zone
column",Y,@ZNSX($Table1.Zone)
```

The production of index values is designed to create an auxiliary attribute table that contains values that can be queried more quickly. Producing an index value for the postal code information as well as the city information will speed the query process. Following the example of a single zone field and considering the functions that are provided within the geocoding software, you may wish to add the following properties:

```
RD.IdxTable1.IdxColumn = Zip,"ZIP code
column",Y,@ZNSX($Table1.Zip)
```

This will perform the same ZNSX indexing function on the information contained in the Zip column.

```
RD.IdxTable1.IdxColumn = City,"City
column",Y,@SDEX($Table1.City)
```

This will perform the SDEX, or Soundex, function to the information in the "City" column. You will also need to assign column names for the index table for these additional columns. This would be done using the following properties:

```
RD.Val.Table1.Zip = ZIP
```

```
RD.Val.Table1.City = CITY
```

Redefining the query process

The final step in accommodating this alternative table schema is to redefine the query process. This process begins by modifying the `RD.<query name>.MF` properties as follows:

```
RD.<query name>.MF = <MAT file defined name> ,
<table name>.<field name>
```

In short, these properties indicate to the geocoding process the address columns in the reference data to use in matching against those similar address elements in the address data. You will notice the reference to the name assigned in the .mat, or match, file. This refers to the names that you assigned when you modified the .mat file by adding the "City" and "ZIP" names as additional search parameters. To accommodate these changes, the property referring to the zone should be removed and replaced by properties referring to the city and postal code. These new properties would appear as such:

```
RD.Query1.MF = City.Table1.City
```

```
RD.Query1.MF = Zip.Table1.Zip
```

Finally, to accommodate the additional indexing values created for the city and postal code information, a modification will need to be made to the "ConditionalWhere" property. This modification would replace the reference to the default "Zone" column with the following:

```
RD.Query1.ConditionalWhere = CT, "AND
($IndTable1.City = '@SDEX(%CT%)')"
```

```
RD.Query1.ConditionalWhere = ZP, "AND
($IndTable1.Zip = '@ZNSX(%ZP%)')"
```

Special cases

The majority of the material already presented in this chapter has dealt specifically with address locators used in ArcGIS Desktop, with customization for reference data stored within a geodatabase. You may find that there are other types of applications that you wish to perform. These may include the use of shapefiles or coverages for your reference data and modifying the locator to be used on an ArcSDE server.

Using shapefiles and coverages as reference data

When using shapefiles and coverages as your reference data in geocoding, there are two major differences from using a feature class within a geodatabase. These differences include the syntax used in the query properties and the method of indexing.

Shapefile indexes

When using a feature class for your reference material, an index table is created that contains a coded value for the information within the other reference data tables. The index table is then linked, along with other related reference material, through the use of related column values. When using shapefiles, the tables cannot be linked together, and the indexing is stored in a binary file. Several groups of binary values are stored within this file, each storing coded index values for specific column attributes from the shapefile. Instead of using a table containing index values as you would when using a feature class, this index file is used to speed the query process.

When creating this file, certain properties differ from those used when creating the index table used with a feature class. They are explained below:

```
RD.IndexFile = <index file name>,${<reference
table name>}
```

```
RD.IndexFile = IdxFile1,${Table1}
```

The initial index-related property specifies the index filename and the reference table name for which the file is being created.

With the index file-related properties, you will also find a property referring to the class ID where the functionality for creating the index file is stored. This property should not be altered and will look as follows:

```
RD.IdxFile1.CLSID = {F48F9A46-F624-11d3-AB54-
00C04FA379E3}
```

The properties that follow define the sections within the index file. This is done through a series of properties that specify the file selection, a common name assigned to it, a function, and the column in the reference material to be used. This appears as such:

```
RD.IdxFile1.Index = <file section where binary
code stored>, "<common name for section>",
<function used to create index values>,
$$<reference table name>.<column name>
```

```
RD.IdxFile1.Index = I1, "Street Name Index",
Soundex2, $$Table1.StreetName
```

Depending on the style you are using, you may find that the house numbers are also indexed. This is needed to specify the specific road segment that contains the house number for which you are searching.

Query properties when using shapefiles

The basic format of the “Where” and “ConditionalWhere” expressions has not changed. However, you will find some differences from those used with a geodatabase. These differences reflect the different indexing techniques and different symbols used in the expression.

Instead of referring to a specific column within the index table, the query properties refer to the specified section of the index file. By default, these sections are given names such as “I1”, “I2”, and so forth. Within the expressions, the section of the file is referred to using these names.

The expressions are also written using different symbols from those used for feature class reference data. The following is an example of one of these properties:

```
RD.<query name>.ConditionalWhere = <match key token>, "expression"
```

RD.Query1.ConditionalWhere = ZN, " & I3 = \"%ZN%\"?"

This "ConditionalWhere" property defines the same principle as described in previous sections; however, many of the words of values are different. Below is a key, defining what each of the symbols means:

Boolean expressions

#	logical NOT
!	alternative for NOT
&	logical AND
/	logical OR
|	alternate for OR
$	logical exclusive OR

Special symbols

?	spelling variations
+	partial key spelling variations
"	treats multiwords as one key
\	partial key separator

ArcSDE address locator files

You can modify a client-side geodatabase address locator file to function within an ArcSDE server. Some modification needs to be made to the file to function correctly. Performing the following edits to the address locator style file will make these needed modifications.

Add the following properties:

```
Category = Address
Name = "<name>"
Description = "<description>"
Type = SE_LOCATOR_TEMPLATE
```

You will also need to remove the following property:

```
RD>Table1.Filter = esriCore.GxFilterPGDBFeatureClasses
```

You will also need to modify the following properties:

```
CLSID = {04FCADCF-ED3B-11D2-9F48-00C04F8ED1C4}
DBEnv.Library = mtchloc
DBEnv.CLSID = {78E9FDDC-001A-11D4-AB58-00C04FA379E3}
```

There are several utilities available to assist in administration of address locator files within ArcSDE. For information related to these utilities, see the *ArcSDE Administration Guide*.

Glossary

actions

Commands in the pattern file that work in conjunction with tokenized addresses to do many things including filtering noise, converting an operand to other values based on a lookup table, copying a value or operand to a match key field or a variable, or invoking a subroutine.

address

A description of a location, most often consisting of specific elements arranged into a particular format.

address data

Material containing the parameters of the location being searched in a geocoding search. The address data may consist of one individual address or a table containing many addresses.

address data format

The arrangement of location-specific information, most often consisting of such address elements as house number, street direction, street name, street type, city, and postal code.

address data model

A geodatabase designed specifically to accommodate address-related material, such as streets, zones, ranges, and so forth. An address data model facilitates address data storage.

address element

A part of an address. A house number, street direction, or street name are all examples of address elements.

address field

A column in a table containing one or some address elements. Address fields can be present in reference data, address data, or both.

address locator

One of the entities in the geocoding framework that acts to combine the style-specific guidelines and reference data. The address locator defines the technique to be used by the geocoding engine in interpreting the address against the rule base files. The file containing information about an address locator is distinguished with a .loc file extension.

Address Locator Properties dialog box

The primary interface in ArcGIS used to create or modify the address locator. The Address Locator Properties dialog box is accessible through both ArcCatalog and ArcMap.

address locator style

A template on which an address locator is built. Each template is designed to accommodate a specific format of address and reference data. The address locator style template file is distinguished with a .lot file extension.

address matching

See geocode.

address standardizer

The address standardizer is what you enter input information into. The input information is found in the first column of the classification table.

address styles

An address style defines the format of addresses and a method of matching that can be used for a specific application.

alternate name

An additional name used to refer to an address feature. In the geocoding service, alternate names are used often for street names. When a street is known by more than one name, such as by a highway number and a street name, the alternate name can be included in the search to replace the related address element.

API (application programming interface)

Application programming interface.

application programming interface (API)

Application programming interface. Refers to a defined and documented set of tools or "functions" that application developers use to build or customize a program or set of programs. APIs can be built for programming languages such as C, COM, Java™, and so on.

ArcSDE

A gateway to a multiuser commercial RDBMS—for example, Oracle®, Microsoft SQL Server™, Informix®, and IBM® DB2®. ArcSDE is an open, high-performance spatial data server that employs client/server architecture to perform efficient spatial operations and manage large, shared geographic data. Was known as a Spatial Database Engine™ (SDE®) before 1999.

arithmetic expressions

Expressions in the pattern file that can be <left-arith-operand>, <arith-operator>, or <right-arith-operand>.

blocking

Occurs during indexing to reduce the number of potential matches that need to be checked.

booster index

A geocoding index designed to reduce the time and resources needed for a geocoding search.

candidate

A location found on the reference data that has the possibility of being a match for the address being searched.

CD

A compact disc, used for data storage.

class

Used in the pattern file to specify the rules that will be used to interpret the elements of an address. A class must be a single character.

classification table

A standard ASCII file in the geocoding rule base that identifies and classifies keywords that may appear in an address, such as street types and directions. The filename is always <file>.cls.

classifying

The process of assigning a class to a part of an address in order to standardize it.

client-side geocoding service

A geocoding service created and used on the same computer.

command file

A file in the geocoding rule base that specifies the standardization commands and processes. The filename is always <file>.stn.

comparison threshold

Degree of uncertainty that can be tolerated in the spelling of the keyword, such as phonetic errors, random insertion, deletion, replacements, and the transposition of characters. The score is weighted by the length of the word, because small errors in long words are less serious than errors in short words.

composite weight

The sum of the individual weights for all field comparisons. The composite weight provides a reference for how good a match is.

CONCAT

A command in the pattern file that concatenates information to a user variable or a match key field. The source can be an operand, a literal, or a user variable.

CONCAT_A

A command in the pattern file that concatenates the standard abbreviation of the operand.

conditional expressions

Expressions in the pattern file that are enclosed in square brackets immediately following the operand. Conditional expressions can include <left-operand>, <relational-operator>, or <right-operand>.

conditional patterns

One type of pattern rule. Conditional pattern rules allow patterns to match only under specified circumstances.

conditional pattern values

Conditional values in patterns can correctly process problem cases, such as ST to SAINT. There are several types of conditional values, including simple conditional values, a series of conditional values, and tables of conditional values. See also simple conditional values, series of conditional values, table of conditional values.

constants

Numeric constants are referenced by coding a number. Negative numbers and decimal points are not permitted in numeric constants.

CONVERT

A command in the pattern file that converts data according to a lookup table instead of a user-supplied literal.

COPY

A command in the pattern file that copies the entire string (all words are concatenated). The form of the command is a <source> and a <target>.

COPY_A

A command in the pattern file that copies the standard abbreviation coded in the classification table for an operand to a target.

COPY_S

A command in the pattern file that preserves spaces between words when you copy them.

DEBUG

Keyword in the command file that puts the standardizer into debugging mode.

dynamic feature class

A feature class consisting of point features associated with the address elements in an address data table that will change based on any changes made to the address data table.

ENCODPAT

One of the tools that comes with STANEDIT in the Geocoding Developer Kit. ENCODPAT is a pattern rule encryption program that is used to encode standardization pattern rule files for use in ESRI's geocoding products. ENCODPAT converts an ASCII .xat file into a binary .pat file.

end of field

A specifier in the pattern file denoted by $. This specifier matches to the end of the field instead of matching to any real token. It is used to ensure that no tokens are left in the field after the pattern.

end offset

A distance away from the end point of a street feature. The end offset is an adjustable value that dictates how far away from the end of a line feature an address location should be matched. Using a side offset will eliminate the point feature from being placed directly over or beyond intersecting line features.

EXIT

A command in the pattern file that quits the pattern matching program for this process in this record, which prevents further pattern–action pairs from being executed.

file

A concatenated method of referring to a shapefile.

fixed position specifier

A specifier in the pattern file denoted by % followed by a numeric value between 0 and 100. It specifies the position at a particular operand in the input string.

floating position specifier

A specifier in the pattern file denoted by *. Used to modify the positioning of the pattern matching. The class immediately following the specifier is a floating class. The pattern is searched until there is a match or until the entire pattern is scanned.

GBF (Geographic Base Files)

See reference files.

GDB

Abbreviation of geodatabase. See geodatabase.

geocode

The process of assigning an x,y coordinate value to the description of a place by comparing the descriptive location-specific elements to those present in the reference data.

Geocoding Editor

Lists the match key fields for the candidate list (with the exception of any identifier that starts with X, for example, XS and XR).

geocoding engine

One of the entities in the geocoding framework that drives the geocoding process.

geocoding index

A table created in conjunction with a geocoding service, designed to decrease the resources necessary to perform a geocoding search. These tables contain a list attributed from the reference attribute table as well as a coded value for the attribute. When geocoding, the same index values are produced for your address query and are matched against the geocoding index.

geocoding process

The steps involved in translating an address entry, searching for the address in the reference data, and delivering the best candidate or candidates. These steps include parsing the address, standardizing abbreviated values, assigning each address element to a category known as a match key, indexing the needed categories, searching the reference data, assigning a score to each potential candidate, filtering the list of candidates based on the minimum match score, and delivering the best match. The process requires reference files, input event records, and software.

geodatabase

An object-oriented geographic database that provides services for managing geographic data. These services include validation rules, relationships, and topological associations. A geodatabase contains feature datasets and is hosted inside a relational database management system.

Geographic Base Files (GBF)

See reference files.

GETMAXSCORE

Calculates the maximum composite score based on a .mat file. You only need this tool for ArcView 3.x.

input event records

Input event record types vary by application. They include customer addresses, location of the event or incident, location of equipment and facilities, and the monument offset.

intersection

The crossing point of two line features. In geocoding, this is most often referring to the crossing point of two streets.

intersection connector

A character used in the address data to indicate that the address consists of two line features. For example, in the address "S. Huntington Dr. & E. Clark Blvd." the ampersand (&) character is the intersection connector. These characters are used to assign intersection searches to specific files in the rule base.

keyword

A single word that appears in the address. A standard abbreviation is substituted for the keyword in the .cls file.

literals

Character constants that are represented by enclosing a string in quotes.

location

A geographic identification assigned to a region or feature based on a specific coordinate system. In the geocoding process, the location is defined with an x,y coordinate value according to the distance north or south of the equator and east or west of the prime meridian.

m probability

The probability that the field agrees, given the record pair is a match. The m probability is one minus the error rate of the field in a matched record. The more reliable a field is, the greater the m probability will be.

MATCH commands

Used to specify the comparison type, match key field, variable name, matching probabilities, and additional parameters when defining match rules for match variables.

match file

A file in the geocoding rule base that defines variables for the address items found in the reference file attribute table. The filename is always <file>.mat.

matching

Performing detailed candidate scoring and field-by-field comparisons in order to find a composite score for each candidate.

match key

A coded element consisting of two characters, used to define and process specific address elements in the geocoding rule base.

match key contents

The match key field name enclosed in braces.

match key dictionary

A file in the geocoding rule base that defines the data type, field length, and missing value code for each match key field. The filename is always <file>.dct.

match key fields

Fields defined in the match key dictionary that contain the data type, field length, and missing value code.

MatchRules

Define variables for the address items found in the reference file attribute table.

match score

A value assigned to all potential candidates of an address match. The match score is based on how well the location found in the reference data matches with the address data being searched.

minimum match score

The minimum score a match candidate needs to be considered in the geocoding search. This value is adjustable on the Address Locator Properties dialog box.

MOVE

A command in the pattern file that moves a user variable or a match key field to another match key field. When you move information using this command, the source is made null after it is moved.

negation class qualifier

A qualifier in the pattern file denoted by "!". This qualifier indicates NOT, and will match to any token except a street type. It can be combined with a floating class only at the beginning of a pattern.

normalized

An internal process (except for ArcView 3.x) where the computed maximum score is recalculated to be between 0 and 100 based on a formula. Normalizing the score allows you to see how good a match is compared to other matches.

operands

A parameter that is referred to in the program. Programming routines or actions can be executed to the specific operand.

OUTFILE

Keyword in the command file that sets the debugging output to <file_name>.

parsing parameters

Default rules in the command file that define what constitutes a token or operand as defined in the pattern file.

pattern–action sequence

Pairs of patterns and actions in the pattern file that work in conjunction with each other. A pattern file can contain as many pattern–action sequences as you want.

pattern classes

Elements in an address are classified using pattern classes in the pattern file. Some examples of pattern classes include the numeric class, single alphabetic word, and unknown. The .cls file also supplies classes A–Z.

pattern file

A file in the geocoding rule base that contains pattern rules and actions for standardizing an address and converting the recognized operands into match key fields. The filename is always <file>.pat.

pattern rules

Rules in the pattern file. There are two types of pattern rules: conditional and unconditional. See also conditional patterns and unconditional patterns.

place name alias

The formal or common name of a location such as the name of a school, hospital, or other landmark. In geocoding, the geocoding service can be designed to accommodate the use of a place name alias as a search parameter.

PREFIX

A command in the pattern file that adds the concatenated operand to the beginning of the string.

PREFIX_A

A command in the pattern file that prefixes the standard abbreviation instead of the raw data. The source must be an operand.

prefix values

Values in an address that are assigned as prefixes by the .cls file. For example, in the address 3453 West Santa Monica Blvd., "West" is the prefix value.

primary reference data

The most basic of reference material used in a geocoding service, usually consisting of the geometry of a region and its associated attribute table.

primary table

The attribute table associated with the primary reference data. Based on the geocoding service style selected, certain address elements must be present in the primary table.

qualifier

A symbol in the pattern file that qualifies the token it is matched with. For example, the ! symbol qualifies the token to be NOT. If the ! is matched to D, and the direction is north, then the qualifier indicates NOT north.

reference data

The material containing location information and address information of specific features. Reference data consists of the geospatial representation of the data, the related attribute table, and the geocoding indexes. The place name alias table and the alternative name table are also considered reference data.

All material contained in the Address data model is used as the reference material in a specific address locator.

reference files

Along with input event records and software, reference files (which are often referred to as Geographic Base Files) are part of the geocoding process. They can vary from simple digitized boundary files to more complex Address Coding Guides to even more sophisticated centerline files.

region

A specified, usually large and continuous, segment of a geographic surface.

reject processing

Performed during the review/edit stage of the geocoding process. Reject processing lets you fine-tune the geocoding process by adjusting index search rules, matching weights, and minimum matching scores.

rematch

The process of reassigning the geocoding results to other locations.

RETURN

A command in the pattern file that returns control from a subroutine to the main program.

RETYPE

A command in the pattern file that lets you change the type of an operand in the token table, change the value of an operand, or change the abbreviation of an operand if it is found in the classification table. You can also retype an operand to a null class, or change the token type and replace the text of a token.

reverse floating position specifier

A specifier in the pattern file denoted by #. This specifier is similar to the floating position specifier, except that scanning proceeds from left to right instead of right to left. See also floating position specifier.

rule base

A collection of files that directs the geocoding engine in how to standardize address data and match it to the related location in the reference data. Each geocoding style uses specific files in the rule base.

SDE

Spatial Database Engine.

secondary reference data

All other material used in a geocoding service as reference data beyond the primary reference data. This can consist of an Alternate Name table or a Place Name Alias table. This entity becomes null when using an address data model as reference material.

SEPLIST

A list of characters or symbols used to define separate tokens. See also parsing parameters.

series of conditional values

Specified by delimiting the conditional values using either spaces or commas.

server-side address locator

An address locator where processing is done on one computer with the results accessible to other computers. Server-side address locators consist of services available over the Internet, via local area networks, or via an ArcSDE server.

shapefile

A vector data storage format for storing the location, shape, and attributes of geographic features. A shapefile is stored in a set of related files and contains one feature class.

side offset

A distance away from the street feature. The side offset is an adjustable value that dictates how far away from the line feature an address location should be matched. Using a side offset will eliminate the point feature from being placed directly over the line feature.

simple conditional values

Denoted by the equality operator and a value. The equality operator tests both the standardized abbreviation and the original token value for equality to the operand.

single token

A pattern class in the pattern file defined by an ampersand.

Soundex

An index search key that finds the match as well as all the potential candidates.

Spatial Database Engine (SDE)

A gateway to a multiuser commercial RDBMS—for example, Oracle, Microsoft SQL Server, Informix, and DB2. ArcSDE is an open, high-performance spatial data server that employs client/server architecture to perform efficient spatial operations and manage large, shared geographic data. SDE has been referred to as ArcSDE since 1999.

specifier

A symbol that specifies what action should be taken with an address token in the pattern file.

spelling sensitivity

The degree to which a geocoding service must match the spelling of one address element in the address data to the address element in the reference data.

standardization

1. A process of address parsing that prepares the address to be matched against the reference data.
2. A two-step process to standardize data. The first step involves preparing the reference file, and the second step involves preparing the events for matching.

standardization process

Contains standardization commands (<file>.stn), a match key dictionary for the event address (<file>.dct), a classification table for providing standard address abbreviations (<file>.cls), and patterns and actions for standardizing an address (<file>.pat). The standardization process can also include additional optional tables.

STANEDIT

The Windows version of Interactive Standardizer. It is used for pattern rules syntax checking and debugging.

STRIPLIST

A list of characters or symbols to be stripped during the standardization. See also parsing parameters.

subfield classes

Classes in the pattern file used to pick off individual words or a ? string. They range from 1 to 9 and -1 to -9. They are useful for processing address suffixes such as 123-A Main St.

subfield ranges

A class in the pattern file used to specify a range of unknown words. Subfield ranges are specified as (beg:end).

subroutines

Actions in the pattern file that are called to perform a particular function (such as CALL_DIRECTIONS, where DIRECTIONS is the subroutine).

table of conditional values

An ASCII file with one line for each conditional value. Tables are specified by @<table file name>.

token

A delimited word, string, or symbol from the entire input string.

tokenizing

The process of assigning a token to a part of an address in order to code it.

token type value

A value assigned to a particular value in the address by the .cls file.

u probability

The probability that the field agrees, given the record pair is unmatched (that is, the probability that the field agrees at random).

unconditional patterns

One type of pattern rule. Unconditional pattern rules are strict rules that are insensitive to the value of the operand.

universal class

A class in the pattern file denoted by **. The universal class matches all tokens. It can be combined with other operands to restrict the tokens grabbed by the class.

user interface

The method of interaction between the user and the software. In the geocoding framework, the user interface consists of ArcCatalog and ArcMap as well as any associated dialog boxes.

VAR commands

Commands in the match file used to specify variable names, field position, field length, and missing value codes.

variables

Can be either numeric or character type. If numeric, its type is set to a numeric value. If character, its type is set to a literal value.

VARTYPE commands

The last line of the match file after you have defined the MatchRules. It is internal, but VARTYPE indicates that frequency analysis won't be performed.

Web service

A software component accessible over the World Wide Web for use in other applications. Web services are built using industry standards such as XML and SOAP and thus are not dependent on any particular operating system or programming language, allowing access through a wide range of applications.

ZIP

1. Zone Improvement Plan. ZIP Code identifies a specific geographic delivery area.
2. The postal code system used by the United States Postal Service.

zone

Additional information about a location or address, used to narrow the geocoding search and increase search speed. Address elements and their related locations, such as city, postal code, or country, all can act as a zone.